THE COMPLETE
COSORI MULTICOOKER ®
COOKBOOK

Quick and Easy Recipes
for Fast and Healthy Meals

Betty Moore

Copyright

TABLE OF CONTENTS

INTRODUCTION

The COSORI Pressure Cooker Multi-Cooker™ is revolutionary! You can cook virtually anything in a pressure cooker — from meats and main courses to rice, potatoes, vegetables of every description, dessert to even yogurt. Better yet, pressure cooking allows you to prepare foods up to 70 percent faster, on average, than conventional cooking methods do, which means you save energy in addition to your precious time!

If you are one of those "hurry-up" cooks who dreams of getting a delicious meal on the table in 30 minutes or less, you'll love this book. I love good meals, but I'm not patient about waiting a long time for it to be done.

Since the pressure cooker came into my life close to ten years ago, I've eaten better and saved money as well as time. Now I'd like to share with you all that I've learned.

In this book, we'll explore the surprising variety of easy dishes you can make with your electric pressure cooker.

For healthy, homemade fast food, the pressure cooker is the best option. We'll explore a wide variety of dishes, from pasta, fish, risottos, meatloafs, and cheesecakes, to all of the splendid soups, stews, ribs, and pot roasts, using wholesome and healthy ingredients in the process.

The pressure cooker makes potatoes creamy and tastier, just as when cooked right, and it does this every time. The pressure cooker works like charm with all potato varieties. In the case of rice, the pressure cooker only needs about ten minutes to make both brown and white rice. What excite me most about it are the rice-based meals such as paella and risotto. I love the Element of Surprise of the electric pressure cooker - you put everything in the pot, close the lid, and let it cook. It isn't until you've released the pressure and opened that lid, until you've dipped a spoon in and had that taste! Is in that moment when you truly know what you have.

I invite you to experience the surprise and delight that awaits every cook who unlocks the lid of a pressure cooker and sees what magic has transpired within.

A pressure cooker is a real kitchen partner.

Let's see what the COSORI Pressure Cooker Multi-Cooker™ can do!

Happy Cooking!

THE REVOLUTIONARY COSORI PRESSURE COOKER MULTI-COOKER™

Fast Food

It greatly decreases typically long cooking times for all dishes. Cooking time can be reduced 60 to 90 percent (depending on the ingredient). Faster cooking times mean you can cook real foods from scratch in the time it takes for pizza delivery or to cook a frozen dinner. A vegetable stock can be ready in 5 minutes; you can cook wholesome soups from scratch in 20 minutes, roasts in 30, braise meat in 40, and desserts in just 20 minutes.

Convenience

COSORI Pressure Cooker Multi-Cooker® features up to ten single key operation buttons to aid in the common tasks of food preparation such as:

- Delay Timer
- Keep Warm
- Meat/Stew
- Poultry
- Soup
- Beans/Chili
- White Rice
- Brown Rice
- Multigrain
- Steam Vegetables
- Slow Cook
- Yogurt

- Pasta Boil
- Steam Potatoes
- Hot Pot
- Sauté/Brown
- Bake
- Reheat
- Time Adjustment
- Keep Warm
- Brown/Sauté
- Stop/Cancel

What's great about the single button operation keys is that they've been skillfully designed to give consistent results with every cooking. And what's more is that if you prefer setting your pressure keeping time according to your recipe, you can achieve it with the manual setting.

Some of the most notable features about the single button operation keys include the following:

Intelligent programming

The programming of the buttons has been done through thousands of trials to equip them with the best features for exceptional cooking results. As in the case of the Rice button for instance, the COSORI Pressure Cooker Multi-Cooker® makes rice and water amount estimates through pre-heating time measurements when cooking rice. Variation of pressure keeping time is achieved based on the measurement. More considerations are done during each of the rice cooking stage including soaking, blanching, steaming and braising.

Refining of each of the function buttons can be done further to vary the taste of the food based on the rare, normal and well-done range.

Automatic cooking

When it comes to a cooking process that's fully automated, then COSORI Pressure Cooker Multi-Cooker® does it all. COSORI Pressure Cooker Multi-Cooker® achieves this by timing every cooking task and then switching to Keep-warm once cooking is through. The convenience with it is that unlike the typical pressure cookers, no stopwatches are needed to monitor cooking time manually.

Planning meals ahead of time

COSORI Pressure Cooker Multi-Cooker® gives the power to plan a meal even before cooking commences. You don't have to spend time around the kitchen watching over the cooker as it does its thing! Making a meal with the COSORI Pressure Cooker Multi-Cooker® gives you all the freedom of being away from the kitchen as the food cooks.

Most importantly, COSORI Pressure Cooker Multi-Cooker® minimizes meal preparation time by 70%. This is particularly a great feature when you have to prepare your dinner in a matter of time after arriving from work in the evening.

Nutritious and healthy meals

COSORI Pressure Cooker Multi-Cooker® seizes the advantages of pressure-cooking by cooking fully flavored, soft and highly nutritious meals. All this is made possible by its cooking cycles, which are controlled by a micro-processor. All meals are cooked with consistency each time.

Retaining food flavor

The COSORI Pressure Cooker Multi-Cooker® confines the food in a completely sealed environment. This traps all the nutrients, aroma and flavor of food rather than dispersing them all over the house. The actual juices of meat, fish, chicken and fruits remain confined in the food hence improving its palatability.

Retaining minerals and vitamins

When steaming food with the COSORI Pressure Cooker Multi-Cooker®, no need for large volume of water, only enough water to generate steam is needed. This is a great feature as minerals and vitamins won't be dissolved and leached away. And since the food is surrounded by steam, oxidation through exposure to air will be absent, so broccoli, asparagus and such vegetables will maintain their brilliant green color as well as phytochemicals.

Soft and tasty meals

A COSORI Pressure Cooker Multi-Cooker® is really great when it comes to preparing tender and tasty meat and bone dishes. Meat is separated from bones completely as the latter become so tender and chewable allowing the body to benefit from calcium among other minerals present in bones. Pressure-cooking also makes beans and whole grain meals so soft and tastier than when cooked with other methods.

One-pot meal.

The electric pressure cooker offers the pleasure of home cooking assembled with little fuss and minimal clean up. With little more to do than toss your ingredients in a pot and turn it on, meals don't get much easier than this. Not everyone has time to prepare three different dishes- a starch, protein and vegetable- every night. The one pot meal allows for diversity, while keeping things simple.

Consistency in results

Consistency in the cooking results is a fully guaranteed in COSORI Pressure Cooker Multi-Cooker®. Pressure-cooking allows even, quick and deep distribution of heat. Intelligent programming of COSORI Pressure Cooker Multi-Cooker® gives guarantee in cooking results regardless of the amount of water and food. This is one unsurpassed advantage that programmable electric pressure cookers have to offer.

Energy Efficiency – Eco-Friendly

When it comes to energy efficiency, COSORI Pressure Cooker Multi-Cooker® is among the most eco-friendly kitchen appliances on the planet. This is so because COSORI Pressure Cooker Multi-Cooker® can save up to 70% of electric power as compared to other cooking appliances such as oven, steamer, boiling pot and stove. Its high-energy efficiency is attributed by the following features:

- We are all aware that food cooks much faster under high pressure and temperature. Reduced cooking time relates to less energy being used. Unlike other cookers, COSORI Pressure Cooker Multi-Cooker® can reduce food cooking time by 70%.

- The exterior housing of the COSORI Pressure Cooker Multi-Cooker® has been insulated completely with two air pocket layers between the exterior and inner pot. This feature concentrates the energy in the inside thus making it more energy efficient as compared to conventional cookers.
- The intelligent monitoring system of COSORI Pressure Cooker Multi-Cooker® heats only the inner pot so as to maintain a particular level of pressure. In longer cooing durations, heating is almost off by 40% of the entire cooking time.

- Since COSORI Pressure Cooker Multi-Cooker® is completely sealed, no water is lost through evaporation and hence less water is needed for cooking. This eventually reduces amount of energy used in meal preparation. Another great advantage of this feature is that the room won't become steamy hot on summers.

Clean and amusing

COSORI Pressure Cooker Multi-Cooker® rattles no unpleasant sounds and is completely quite during operation, unlike the conventional pressure cookers. The inner pot is completely sealed even when pressure build up becomes high. No steam and aroma will escape hence your house will be free from smell. There won't be messy spills, spatters and splashes. Your room does not heat up and this is pretty much appreciated during the summers. This is indeed a friendly kitchen appliance to have. And since it's a multi-cooker, it reduces number of cooking appliances in the kitchen, and you get to save yourself cash and reducing clutter hence an organized kitchen.

TESTING AND USING THE COSORI PRESSURE COOKER MULTI-COOKER®

Initial Test Run

In order to familiarize yourself with your COSORI Pressure Cooker Multi-Cooker ® and start cooking with your favorite recipe, it is highly recommended for first time users to do a test run. This test run will help you become familiar with the COSORI Pressure Cooker Multi-Cooker ®, make sure the unit is free of residues from the manufacturing process, and indicate whether your unit is working properly. This optional, but recommended, test run will complete in about 15 minutes.

1. Make sure the pressure release handle and float valve are unobstructed and clean, and that the sealing ring is properly inserted.
2. Place the inner pot in the cooker base and add 3 measuring cups of water into the inner pot.
3. Close the lid. Make sure that the pressure release handle is close.
4. Press the "Steam/Vegetables" button, 3 minute cook time.
5. In 10 seconds, your COSORI Pressure Cooker Multi-Cooker ® will go into the preheating cycle. Within a few minutes, steam will release for a minute or two until the Float Valve pops up to seal the cooker. Once working pressure is reached, which may take a few minutes or up to 10 -13 minutes, the countdown timer will begin. When the countdown is finished, the COSORI Pressure Cooker Multi-Cooker ® will beep and automatically switch into the "Keep Warm" mode.

The test is now complete. Press the "Stop/Cancel" button and/or unplug the unit. Once the pot cools down, you can open it and are ready to try your favorite recipes.

Electric Pressure Cooking – Four Steps

Pressure cookers can save you time and money, helping you prepare delicious meals that retain nutritional values often lost in other cooking methods.

If you're new to pressure cooking, It is very important to read your pressure cooker's manual so you understand how your specific model works: the operating pressure, the type of pressure signal, how to release the pressure when cooking time is up, how to remove the pressure valve for cleaning, etc.

The electric pressure cooker is very simple; after you've read the manual, let's go over the basics of cooking in a pressure cooker.

Classic electric pressure-cooking recipes will involve the following steps.

1. **Preparing your ingredients (Brown Meat, then Aromatics)**

Prepare ingredients according to the directions in the pressure-cooking recipe you have selected. For extra flavor, use the brown or sauté functions first, just like you would when cooking with conventional cookware. For instance, brown the meat and vegetables for a stew, before adding other liquids and cooking under pressure. Be sure to deglaze the pot, scraping up any browned bits clinging to the bottom with a small amount of wine, broth or even water, so they are loosen, adding flavor to your food, as well as discouraging scorching.

2. **Add Liquid**

After the aromatics are softened, add the remaining ingredients and pour liquid, into the cooker body, as specified in the recipe or timetable. This liquid is usually water. However, some recipes will call for other liquids, such as wine.

3. **Lock the lid and select the function**

Once you have all the ingredients in the cooker, lock the lid in place, and select the function, according to the recipe. The cooker will automatically start.

4. **Turn off the cooker and release the pressure**

After the pressure-cooking time has finished, turn off the cooker by selecting "Stop/Cancel" button. You can release the pressure two ways: quick release and natural release, according to the recipe or timetable instructions.

A **quick pressure release** is when you open the valve and allow the steam to release quickly. The pressure cooker will not open until the pressure is released and the float valve has dropped.

A **natural pressure release** is when the cooking time is over and you leave the valve closed and allow the pressure to decrease without doing anything. A natural pressure release can take from 5 to 30 minutes in the electric pressure cooker. When the pressure is fully released, the float valve will drop and the lid will unlock and open.

5. **Finish the dish**

In some cases, after releasing pressure and carefully removing the lid, some dishes need from a simmer to help thicken, reduce, or concentrate the liquid; others require to add more ingredients to finish the recipe.

THE MAGIC OF THE COSORI PRESSURE COOKER MULTI-COOKER ® (MULTI-COOKER)

That you know the Magic COSORI Pressure Cooker Multi-Cooker ® is a multi-cooker that works as an electric pressure cooker, slow cooker, rice cooker, steamer, warmer, and sauté pan all in one.

Depending on the brand and model you purchase, they can perform up to 10 different functions in the kitchen. I'll start with the buttons I use the most.

1. **Stop/Cancel** – Use this button to cancel a function or turn off your pressure cooker. When you pressure cooking time is up, it will automatically switch to Keep Warm.
2. **Delay Start Timer** – allow you to set the pressure cooker to start cooking later in the day.
3. **Slow Cook** – 2 hour cook time, use the cook time selector to adjust to 6 hours or 12 hours.

If you prefer to use the pre-set buttons instead of the manual button, here's a concise description of what the pressure-cooking buttons/ functions do:

4. **Steam** – 10 minute cook time, use the cook time selector to adjust to 3 or 60 minutes. This is the shortest cook time available on the COSORI Pressure Cooker Multi-Cooker
5. **Beans/Chili** – 20 minute cook time, use the cook time selector to adjust to 5 minutes or 2 hours.
6. **Rice** – 12 minute cook time; use the cook time selector to adjust to 6 or 30 minutes. The manual suggests using 6 minutes for white rice, 18 minutes for brown rice, and 25 minutes for wild rice.
7. **Soup** – 30 minute cook time; use the cook time selector to adjust to 5 minutes or 2 hours.

8. **Meat/Stew** – 35 minute cook time; use the cook time selector to adjust to 15 minutes or 2 hours.
9. **Poultry** –15 minute cook time; use the cook time selector to adjust to 15 minutes or 2 hours.
10. **Multigrain** -40 minute cook time; use the cook time selector to adjust to 10 minutes or 2 hours.
11. **Yogurt** - 8 hours low temp cook time ; use cook time selector to adjust to 6 or 12 hours.

.

TIPS FOR NEWBIES

1. When you start, the pressure valve on top must be set to Locked position, not open. This means that the pot is sealed shut and steam cannot come out. Therefore, the pressure can build up.
2. Then make sure the lid is closed tightly. Before you close it, make sure that the white silicone ring on the inside of the lid is in place all the way around.
3. Choose which button you want to push, depending on your recipe and push it.
4. Next, the number of minutes that your thing will cook will show up and it will start counting down.
5. When it gets to "0", you can leave it alone and let it "NPR" or "Naturally Release the Pressure", for as long as you like or as long as your recipe specifies. During this time, the numbers will go back up and these numbers tell you how many minutes it has been in the "NPR" state. The metal pin will drop down into the lower position when it is safe to open it.
6. If you want to let the pressure out quickly, right when it's done or any time later on, turn the pressure valve to "Open Position" and get your hands out of the way. Steam will come shooting out of the float valve. When it stops shooting out and the metal pin drops down to the lower position, it is safe to open.
7. The "timer" button is NOT a timer button! It is really a "delay start" button and is only used for setting the amount of delay you want before the cooking starts.
8. If you have a recipe for unfrozen meat, and your meat is frozen, use the same cooking time and amount of liquid indicated in your recipe. The difference is that the time for the COSORI Pressure Cooker Multi-Cooker ® to come to pressure will increase.

Cinnamon Honey Applesauce

PREP: 5 MINUTES • PRESSURE: 6 MINUTES • TOTAL: 11 MINUTES • PRESSURE LEVEL: HIGH • RELEASE: NATURAL
SERVES: 6-8

Ingredients

3 pounds medium-tart baking apples, such as McIntosh, cored, peeled, and roughly chopped
¾ cup unsweetened apple juice
⅓ cup honey
1-tablespoon fresh lemon juice
½ teaspoon ground cinnamon
½ teaspoon salt

Directions

1. **Preparing the Ingredients**. Mix everything in the COSORI Pressure Cooker Multi-Cooker ®.
2. **High pressure for 6 Minutes**. Lock the lid onto the pot. Set the COSORI Pressure Cooker Multi-Cooker to cook at high pressure for 6 minutes, using the "White Rice" Button and use the TIME ADJUSTMENT button to adjust the cook time to 6 minutes. Turn off the COSORI Pressure Cooker Multi-Cooker ® or unplug it so it doesn't flip to its keep-warm setting.
3. **Pressure Release.** Reduce the pressure with Natural Release Method. Unlock and open the COSORI Pressure Cooker Multi-Cooker.
4. **Finish the dish.** Use an immersion blender or a potato masher right in the pot to puree the apples into a thick sauce.
5. Serve warm and enjoy!

Apples And Cinnamon Oatmeal

PREP: 5 MINUTES • PRESSURE: 12 MINUTES • TOTAL: 17 MINUTES • PRESSURE LEVEL: HIGH • RELEASE: NATURAL

SERVES: 4

Ingredients

3 cups water
2 tablespoons packed brown sugar
½ teaspoon ground cinnamon
¼ teaspoon kosher salt
¾ cup steel-cut oats
1 small apple, peeled, cored, and diced
1 teaspoon unsalted butter
1-tablespoon heavy (whipping) cream

Directions

1. **Preparing the Ingredients.** In the COSORI Pressure Cooker Multi-Cooker®, stir together the water, brown sugar, cinnamon, and kosher salt, dissolving the salt and sugar. Pour in the oats, add the apple, and stir again.
2. **High pressure for 12 minutes.** Lock the lid in place, cook for 12 minutes. To get 12 minutes cook time, press the "Manual" button and adjust the time to 12 minutes. When the time is up turn the COSORI Pressure Cooker Multi-Cooker off. ("Keep warm" setting, turn off).
3. **Pressure Release.** Use the natural release method. Unlock and open the COSORI Pressure Cooker Multi-Cooker ®.
4. **Finish the dish.** Stir the oats, and taste; if you like them softer, place the lid on the cooker, but *don't lock* it. Let the oats sit for 5 to 10 minutes more. When they are ready to serve, stir in the butter and heavy cream.
5. Serve and Enjoy!

PER SERVING: CALORIES: 181; FAT: 4G; SODIUM: 157MG; CARBOHYDRATES: 31G; FIBER: 4G; PROTEIN: 5G

Apple, bacon And Grits Casserole

PREP: 15 MINUTES • PRESSURE: 20 MINUTES • TOTAL: 35 MINUTES • PRESSURE LEVEL: NORMAL • RELEASE: QUICK

SERVES: 4-6

Ingredients

2 tablespoons unsalted butter, plus more for buttering the dish

8 ounces Canadian bacon, chopped

1 medium tart green apple, such as Granny Smith, peeled, cored, and chopped

4 medium scallions, green and white parts, trimmed and sliced into thin bits

1 teaspoon dried thyme

¾ cup quick cooking or instant grits

2 large eggs, lightly beaten

½ cup shredded Cheddar cheese (about 2 ounces)

Directions

1. **Preparing the Ingredients.** Melt the butter in the COSORI Pressure Cooker Multi-Cooker ® turned to the Browning function. Add the Canadian bacon; cook, stirring often, for 1 minute. Add the apple, scallions, and thyme; cook for 1 minute, stirring constantly. Scrape the contents of the cooker into a large bowl. Wipe out the cooker with a damp paper towel.
    Turn the COSORI Pressure Cooker Multi-Cooker® to its Browning mode. Add 3 cups water and bring to a boil. Whisk in the grits and cook, whisking all the while, until thickened, about 5 minutes. Scrape the grits into the bowl with the bacon mixture; cool for 10 minutes. Wash and dry the cooker.
    Set the rack inside the cooker and pour in 2 cups water. Make a foil sling and set a 2-quart, high-sided, round baking or soufflé dish on top of it. Lightly butter the inside of the dish.
    Stir the eggs and cheese into the grits mixture until uniform and well combined. Spread the mixture in the prepared baking dish; cover and seal with foil. Lower the dish onto the rack in the cooker with the sling. Fold the ends of the sling so they'll fit inside the cooker.
2. **High pressure for 22 minutes.** Lock the lid onto the pot and cook at high pressure for 22 minutes. To get 22 minutes cook time, press the ""Manual"" button and then use the"time adjustment" button to adjust to 22 minutes.
3. **Pressure Release.**Use the quick-release method.
4. **Finish the dish**. Unlock and open the cooker. Use the sling to transfer the baking dish to a wire cooling rack, steadying the dish as necessary. Uncover, cool a couple of minutes, and spoon the casserole onto individual plates to serve.

Banana Oatmeal

PREP: 5 MINUTES • PRESSURE: 18 MINUTES • TOTAL: 24 MINUTES • PRESSURE LEVEL: HIGH • RELEASE: NATURAL

SERVES: 4

Ingredients

½ cup steel-cut oats
½ cup packed light brown sugar
2 ripe bananas, chopped
2 teaspoons vanilla extract
½ teaspoon ground cinnamon
¼ teaspoon salt
¼ cup heavy cream

Directions

1. **Preparing the Ingredients.** Mix the oats, brown sugar, bananas, vanilla, cinnamon, and salt with 2¼ cups water in the COSORI Pressure Cooker Multi-Cooker® until the brown sugar dissolves.
2. **High pressure for 18 minutes**. Lock the lid onto the pot and cook at high pressure for 18 minutes. To get 18 minutes cook time, press the ""Manual"" button and use the TIME ADJUSTMENT button to adjust the cook time to18 minutes.
3. **Pressure Release.** Turn off the COSORI Pressure Cooker Multi-Cooker® or unplug it so it doesn't flip to its keep-warm setting. Allow the pot's pressure to come to normal naturally, 10 to 12 minutes.
   If the pot's pressure hasn't returned to normal within 12 minutes, use the quick-release method to bring it back to normal.
4. **Finish the dish** Unlock and open the cooker. Stir in the cream and set aside for 1 minute to warm before serving.

Egg and Cheese Breakfast

PREP: 5 MINUTES • PRESSURE: 4 MINUTES • TOTAL: 9 MINUTES • PRESSURE LEVEL: HIGH • RELEASE: QUICK
SERVES: 2

Ingredients

1 teaspoon-unsalted butter, at room temperature, divided
2 large eggs
¼ teaspoon kosher salt, divided
Freshly ground black pepper
2 tablespoons grated aged Cheddar or Parmesan cheese, divided
1-cup water, for steaming
2 English muffins

Directions

1. **Preparing the Ingredients.** Using ½ teaspoon of butter each, coat the insides of 2 heatproof custard cups or small ramekins. Crack 1 egg into each cup, and carefully pierce the yolks in several places to make sure the yolk cooks through evenly. Sprinkle each with ⅛ teaspoon of kosher salt, some pepper, and 1 tablespoon of Cheddar cheese, covering the eggs. Cover the cups with aluminum foil, crimping it around the sides.
Add water and insert the steamer basket or trivet. Place the cups on the insert.
2. **High pressure for 4 minutes.** Lock the lid in place, and bring the pot to high pressure for 4 minutes. To get 4-minutes cook time, press the "Manual" button and use the COOK TIME SELECTOR button to adjust the cook time to 4 minutes.
3. **Pressure Release** After the timer reaches 0, the cooker will automatically enter Keep warm mode. Press the Stop button and carefully release the pressure.
4. **Finish the dish** Toast the English muffins while the eggs cook.
Unlock but *don't remove* the lid for another 30 seconds; this helps ensure that the whites are fully cooked. Using tongs, remove the cups from the cooker and peel off the foil.
Using a small offset spatula or knife, loosen the eggs, then tip each one out onto the bottom half of one of the English muffins.
5. Top with the other half, and enjoy.

PER SERVING: CALORIES: 241; FAT: 9G; SODIUM: 682MG; CARBOHYDRATES: 26G; FIBER: 2G; PROTEIN: 14G

Quinoa Breakfast Bowl

PREP: 15 MINUTES • PRESSURE: 7 MINUTES • TOTAL: 22 MINUTES • PRESSURE LEVEL: HIGH • RELEASE: NATURAL
SERVES: 4

Ingredients
1-cup (173 g) quinoa
1 1/2 cups (350 ml) water
3/4 teaspoon kosher salt, divided
1-pint cherry tomatoes (25 to 30 tomatoes)
1-tablespoon (15 ml) extra-virgin olive oil
1/4 teaspoon freshly ground black pepper
2 scallions (white and light green parts), thinly sliced
2 tablespoons (8 g) chopped fresh flat-leaf parsley
1 avocado
2 large eggs, hard-boiled, cooled, and peeled

Directions

1. **Preparing the Ingredients** Using a fine-mesh strainer, rinse the quinoa, then place into the COSORI Pressure Cooker Multi-Cooker®. Add the water and 1/2 teaspoon of the salt.
2. **High pressure for 7 minutes.** Close the lid and Cook for 7 minutes. To get 7-minutes cook time, press the "Manual" button and use the TIME ADJUSTMENT button to adjust the cook time to 7 minutes.
3. **Pressure Release** Use the "Natural Release" method for 5 minutes, then vent any remaining steam and open the lid.
    Fluff with a fork. Press On/Start, lock the lid, and let sit for 5 minutes more.
4. **Finish the dish** While the quinoa is cooking, preheat broiler. On a small rimmed baking sheet, toss the tomatoes with the olive oil, pepper, and the remaining 1/4 teaspoon salt. Broil until the tomatoes begin to burst, about 3 minutes. Toss with the scallions and parsley.
    Pit, peel, and dice the avocado. Divide the quinoa among bowls, top with the tomatoes and avocado, and then coarsely grate the eggs on top.
5. Serve and Enjoy!

"Softboiled" Eggs

PREP: 5 MINUTES • PRESSURE: 3 MINUTES • TOTAL: 8 MINUTES • PRESSURE LEVEL: HIGH • RELEASE: QUICK
SERVES: 2

Ingredients

2 teaspoons unsalted butter, at room temperature, divided

2 large eggs

¼ teaspoon kosher salt, divided

Freshly ground black pepper

1-cup water, for steaming

2 slices of toast (optional)

Directions

1. **Preparing the Ingredients** Using ½ teaspoon of butter each, coat the insides of 2 heatproof custard cups or small ramekins. Crack 1 egg into each cup, and sprinkle each with ⅛ teaspoon of kosher salt and some pepper. Divide the remaining 1teaspoon of butter in half, and top each egg with one piece. (You can omit the butter on top of the egg, but it is delicious. Don't skip buttering the dish, though, or the egg won't come out.) Cover the cups with aluminum foil, crimping it down around the sides.
Add the water and insert the steamer basket or trivet. Carefully transfer the cups to the steamer insert.
2. **High pressure for 3 minutes** Close the lid and the pressure valve and then cook for 3 minutes. To get 3-minutes cook time, press "MANUAL" button and use the TIME ADJUSTMENT button to adjust the cook time to 3 minutes.
3. **Pressure Release** Use the quick-release method.
4. **Finish the dish** Unlock but *don't remove* the lid for another 30 seconds; this will help ensure that the whites are fully cooked. Using tongs, remove the cups from the cooker and peel off the foil. Scoop each egg out onto a slice of toast (if desired).
5. Serve and Enjoy!

PER SERVING: CALORIES: 105; FAT: 9G; SODIUM: 388MG; CARBOHYDRATES: 0G; FIBER: 0G; PROTEIN: 3G

Mini Fritatas

PREP: 5 MINUTES • PRESSURE: 5 MINUTES • TOTAL: 10 MINUTES • PRESSURE LEVEL: NORMAL • RELEASE: QUICK
SERVES: 5

Ingredients

5 eggs
Splash of milk (I use almond milk)
Spices such as salt and pepper
Desired mix in's: cheese, veggies, meats, the options are endless!

Directions

1. **Preparing the Ingredients** .Mix eggs, milk, and mix-in's in a dish.
   Pour mixture into individual baking molds, I use silicone molds.
   Place molds on rack in COSORI Pressure Cooker Multi-Cooker ® with 1 cup of water.
2. **High pressure for 5 minutes**. Close the lid and the pressure valve and then cook for 5 minutes. To get 5-minutes cook time, press "Manual" button and use the TIME ADJUSTMENT button to adjust the cook time to 5 minutes.
3. **Pressure Release** Use the quick-release method when the timer goes off and cooking is done.
4. Enjoy!

Cinnamon, Vanilla and Banana Bread

PREP: 5 MINUTES • PRESSURE: 40 MINUTES • TOTAL: 45 MINUTES • PRESSURE LEVEL: HIGH • RELEASE: QUICK
SERVES: 6

Ingredients

1/4 cup (1/2 stick, or 60 g) unsalted butter, melted, plus more for the pan
1 cup (120 g) all-purpose flour
1/2 teaspoon baking powder
1/4 teaspoon baking soda
1/2 teaspoon kosher salt
1/8 teaspoon ground cinnamon, plus more for dusting
2 large eggs
1/3 cup (65 g) sugar
1/3 cup (77 g) sour cream
1/2 teaspoon pure vanilla extract
1 large ripe banana, mashed
1/2 cup (55 g) pecans, chopped

Directions

1. **Preparing the Ingredients.** Insert the steam rack into the COSORI Pressure Cooker Multi-Cooker® and add 1 1/2 cups (350 ml) water. Butter a 6 × 3-inch (15 × 7.5 cm) round cake pan.
   In a medium bowl, whisk together the flour, baking powder, baking soda, salt, and cinnamon.
   In a second medium bowl, whisk together the eggs, sugar, sour cream, melted butter, and vanilla. Mix in the banana. Add the dry ingredients and mix to combine; stir in the pecans.
   Scrape the batter into the prepared pan and cover with aluminum foil. Place the pan on the steam rack. Lock the lid.
2. **High pressure for 40 minutes.** Close the lid and the pressure valve and then cook for 40 minutes. To get 40 minutes cook time, press "Manual" button and use the TIME ADJUSTMENT button to adjust the cook time to 40 minutes.
3. **Pressure Release** .Use the "Quick Release" method.
4. Serve and Enjoy!

Chapter Seven

Main Dishes – Beef and Lamb

Classic Pot Roast

PREP: 5 MINUTES • PRESSURE: 90 MINUTES • TOTAL: 95 MINUTES • PRESSURE LEVEL: HIGH • RELEASE: QUICK AND NATURAL

SERVES: 6

Ingredients

1 tablespoon olive oil
One 3- to 3½-pound boneless beef chuck roast
1 teaspoon salt
½ teaspoon ground black pepper
1 large yellow onion, chopped
2 teaspoons minced garlic
Up to 1½ cups beef broth
3 tablespoons tomato paste
One 4-inch rosemary sprig
½ ounce dried mushrooms, preferably porcini
1½ pounds small white or yellow potatoes

Directions

1. **Preparing the Ingredients** Heat the oil in the COSORI Pressure Cooker Multi-Cooker®. Turn on the pressure cooker to the Sauté setting then wait for it to boil.
   Season the roast with the salt and pepper; brown it on both sides, turning once, about 10 minutes. Transfer the meat to a large bowl.
   Add the onion; cook, stirring often, until translucent, about 4 minutes. Add the garlic; cook, stirring constantly, until aromatic, about 30 seconds. Pour 1¼ cup broth in the COSORI Pressure Cooker Multi-Cooker®. Add the tomato paste and stir well until dissolved. Tuck the rosemary into the sauce and crumble in the mushrooms. Nestle the meat into the sauce, adding any juices in the bowl.
2. **High pressure for 60 minutes.** Close the lid and the pressure valve and then cook for 60 minutes. To get 60-minutes cook time, press "Manual" button and use the TIME ADJUSTMENT button to adjust the cook time to 60 minutes.
3. **Pressure Release** Use the quick-release method.
   Unlock and open the cooker; sprinkle the potatoes around the meat.
4. **High pressure for 30 minutes.** Close the lid and the pressure valve again and cook for 30 minutes. To get 30-minutes cook time, press "Manual" Button.
5. **Pressure Release** Use the natural-release method -20 to 30 minutes.

6. **Finish the dish.** Transfer the roast to a cutting board; set aside for 5 minutes. Discard the rosemary sprig. Slice the meat into 2-inch irregular chunks and serve these in bowls with the vegetables, mushrooms, and broth.
7. Serve and Enjoy!

Brisket With Veggies

PREP: 10 MINUTES • PRESSURE: 60 MINUTES • TOTAL: 70 MINUTES • PRESSURE LEVEL: HIGH • RELEASE: QUICK
SERVES 6

Ingredients

2 tbs. olive oil
5 or 6 red potatoes
2 lb. or larger regular brisket, rinsed and patted dry
Fresh ground black pepper
3 tbs. heaping chopped garlic
1 lg. yellow onion
2 c. large chunks carrots
2-½ c. home made beef broth, or make from Knorr Beef Base
3 tbs. Worcestershire Sauce
4 bay leaves
5 or 6 red potatoes
Granulated garlic
Knorr Demi-Glace sauce
½ c. dehydrated onion
2 stalks celery in 1" chunks

Directions

1. **Preparing the Ingredients** Put the COSORI Pressure Cooker Multi-Cooker® on the sauté setting.
   Put in 1 tbs. (more if needed) of the oil and caramelize the onions. Once golden, remove from pot, put in a bowl, and set aside. But keep the COSORI Pressure Cooker Multi-Cooker® on the "Sauté" setting.
   Rub the freshly ground pepper on both sides of the brisket. Do the same with the granulated garlic. Add 1tbs. olive oil (or more) and only lightly sear the brisket on all sides.
   Add back the onions, garlic, Worcestershire sauce, bay leaves, dehydrated onion and beef broth.
2. **High pressure for 50 minutes**. Close the lid and the pressure valve and then cook for 50 minutes. To get 50-minutes cook time, press "Manual" button and use the TIME ADJUSTMENT button to adjust the cook time to 50 minutes.
   While the meat is cooking, peal and cut up all the veggies. When the meat is done, use the quick pressure release feature, and then remove the lid. Add all of the veggies, replace the lid and cook at high pressure for to 10 minutes. To get 10-minutes cook time, press Steam button
3. **Pressure Release** When the time is up, turn the pot off, use the quick release again, and remove the lid.
4. **Finish the dish.** Use a platter to remove the veggies and meat. Use the "Sauté" setting and bring the broth to a boil, then add the Knorr Demi-Glace mixing with a Wisk. Adjust seasonings as needed. Serve with Cole Slaw or other salad, home made rolls or Italian garlic bread. Be sure to remove the bay leaves before serving.

5. Serve and Enjoy

Per Serving Calories: 425; Total Carbohydrates: 50g; Saturated Fat: 3.6g; Trans Fat: 0g; Fiber: 10.6g; Protein: 30.5g; Sodium: 490mg

Korean Braised Short Ribs

PREP: 10 MINUTES • PRESSURE: 45 MINUTES • TOTAL: 55 MINUTES • PRESSURE LEVEL: HIGH • RELEASE: NATURAL

SERVES 4-6

Ingredients

1 teaspoon vegetable oil
2 green onions cut into 1-inch lengths
3 cloves garlic, smashed
3 quarter-sized slices of ginger
4 pounds beef short ribs, about 3 inches thick, cut into 3 rib portions
1/2-cup water
1/2-cup soy sauce
1/4-cup rice wine (or dry sherry)
1/4-cup pear juice (or apple juice)
2 teaspoons sesame oil
Minced green onions
Gochujang sauce

Directions

1. **Preparing the Ingredients** Heat the vegetable oil in the COSORI Pressure Cooker Multi-Cooker® using the "Sauté" function, until the oil is shimmering. Add the green onion, garlic, and ginger, and sauté for 1 minute, or until you can smell garlic. Add the short ribs, water, soy sauce, rice wine, pear juice and sesame oil. Stir until the ribs are completely coated.
2. **High pressure for 45 minutes**. Lock the lid on the COSORI Pressure Cooker Multi-Cooker® and then cook for 45 minutes. To get 45-minutes cook time, press Meat/Stew button and use the ADJUST button to adjust the cook time to 45 minutes.
3. **Pressure Release** Let the pressure to come down naturally for at least 15 minutes, then quick release any pressure left in the pot.
4. **Finish the dish** Remove the short ribs from the pot with a slotted spoon.
5. Serve the ribs with the degreased sauce.

Thai Red Beef Curry

PREP: 15 MINUTES • PRESSURE: 45 MINUTES • TOTAL: 60 MINUTES • PRESSURE LEVEL: MEDIUM • RELEASE: NATURAL

SERVES 6-8

Ingredients

1 tablespoon vegetable oil

1 medium onion, peeled, and sliced into 1/2 inch wedges

1 red bell pepper, cored, stemmed, and sliced into 1/2 inch strips

3 cloves garlic, crushed

1/2 inch piece of ginger, peeled and crushed

Cream from the top of a (13.5 ounce) can coconut milk

4 tablespoons red curry paste (a whole 4 oz. can)

8-ounce can bamboo shoots, drained

2 pounds flat iron steak (or chuck blade steak), cut into 2 inch by 1/2 inch strips

1 teaspoon Diamond Crystal kosher salt or 2 teaspoons fine sea salt

1/2 cup chicken stock or water

1 tablespoon fish sauce (plus more to taste)

1 tablespoon soy sauce (plus more to taste)

Juice from 1 lime

Minced cilantro

Minced basil (preferably Thai basil)

Lime wedges

Jasmine rice

Directions

1. **Preparing the Ingredients.** Heat the vegetable oil in the COSORI Pressure Cooker Multi-Cooker® using the "Sauté" function, until the oil is shimmering. Stir in the onion, red bell pepper, garlic, and ginger, and sauté until the onion starts to soften, about 3 minutes.

   Fry the curry paste: Scoop the cream from the top of the can of coconut milk and add it to the pot, then stir in the curry paste. Cook, stirring often, until the curry paste darkens, about 5 minutes.

   Sprinkle the beef with the kosher salt. Add the beef to the pot, and stir to coat with curry paste. Stir in the rest of the can of coconut milk, bamboo shoots, chicken stock, fish sauce, and soy sauce.

2. **High pressure for 12 minutes**. Lock the lid on the COSORI Pressure Cooker Multi-Cooker® and then cook for 12 minutes. To get 12-minutes cook time, press "Manual" button and adjust the cook time.

3. **Pressure Release.** Let the pressure to come down naturally for at least 20 minutes, then quick release any pressure left in the pot.

4. **Finish the dish.** Remove the lid from the COSORI Pressure Cooker Multi-Cooker®. Stir in the lime juice, and then taste the curry for seasoning, adding more fish sauce

or brown sugar as needed. Ladle the curry into bowls, sprinkle with minced cilantro and basil, and serve with Jasmine rice.

5. Serve and Enjoy!

Per Serving Calories: 321.5; Protein: 28.8g

Beef Ribs

PREP: 10 MINUTES • PRESSURE: 60 MINUTES • TOTAL: 70 MINUTES • PRESSURE LEVEL: HIGH • RELEASE: NORMAL

SERVES 4-6

Ingredients

1 tablespoon sesame oil
2 cloves garlic, peeled and smashed
1" knob fresh ginger, peeled and finely chopped
1 pinch red pepper flakes
¼ cup rice vinegar (or white balsamic vinegar)
⅓ cup raw sugar
⅔ cup soy sauce
⅔ cup salt-free (home made) beef stock
4 pounds (2k) beef ribs (about 8), ask butcher to saw or chop them in half
2 tablespoons cornstarch
1-2 tablespoons water

Directions

1. **Preparing the Ingredients** Turn on the COSORI Pressure Cooker Multi-Cooker® to "Sauté" mode.
   Add sesame oil garlic, ginger and red pepper flakes and sauté for a minute.
   Then, de-glaze with vinegar, mix-in the sugar, soy sauce and beef stock - mix well.
   Add the ribs to the COSORI Pressure Cooker Multi-Cooker® coating them with the mixture.
2. **High pressure for 60 minutes.** Close and lock the lid of the COSORI Pressure Cooker Multi-Cooker®, cook at high pressure for 60 minutes. To get 60-minutes cook time, press "Manual" button and use the TIME ADJUSTMENT button to adjust the cook time to 60 minutes.
3. **Pressure Release** Use the Natural release method (20 minutes).
4. **Finish the dish** Remove the ribs, and place on a cookie sheet and slide under the broiler for about 5 minutes to brown. Make a slurry with the corn starch and water and then mix into the rib cooking liquid in the COSORI Pressure Cooker Multi-Cooker®."Sauté" the mixture until it reaches the desired consistency.
5. Serve and Enjoy!

Per Serving Calories: 307.3; Carbohydrates: 8.6g; Fat: 10.7g; Fiber: 10.6g; Protein: 32.3g; Sodium: 1654.6mg; Cholesterol: 89.2g

Lamb Casserole

PREP: 15 MINUTES • PRESSURE: 35 MINUTES • TOTAL: 50 MINUTES • PRESSURE LEVEL: HIGH • RELEASE: NORMAL

SERVES 6-8

Ingredients

- 1 pound of baby potatoes
- 1 pound rack of lamb
- 2 carrots
- 1 large onion
- 2 stalks of celery
- 1-2 teaspoons of salt depending on the salt content of the chicken stock
- 2 medium size tomatoes
- 2 cups of chicken stock
- 3-4 large cloves of garlic
- 2 teaspoon of cumin powder
- 2 teaspoon of Paprika
- A pinch of dried rosemary
- A pinch of dried oregano leaves
- 2 table spoons of ketchup
- 3 table spoons of sherry or red wine
- A splash of beer if you have one in hand

Directions

1. **Preparing the Ingredients** Dice the tomatoes, onion and garlic, cut potatoes and carrots, cut the rack of lamb to two halves. Put all the ingredients, in the COSORI Pressure Cooker Multi-Cooker®.
2. **High pressure for 35 minutes.** Lock the lid on the COSORI Pressure Cooker Multi-Cooker® and then cook for 35 minutes. To get 35-minutes cook time, press "Manual" button.
3. **Pressure Release** Use Natural-Release Method for 10 minutes, and then Quick-Release.
4. Serve and Enjoy!

Per Serving Calories: 407.3; Carbohydrates: 6.6g; Fat: 11.7g; Fiber: 8.6g; Protein: 35.3g; Sodium: 1640.6mg; Cholesterol: 77.2

Barbecued Baby Back Ribs

PREP: 5 MINUTES • PRESSURE: 32 MINUTES • TOTAL: 37 MINUTES • PRESSURE LEVEL: HIGH • RELEASE: NATURAL
SERVES 4

Ingredients
¼ cup canned tomato paste
2 tablespoons cider vinegar
1 tablespoon sweet paprika
½ tablespoon coriander seeds
½ tablespoon fennel seeds
1 teaspoon onion powder
1 teaspoon dried thyme
½ teaspoon ground allspice
½ teaspoon salt
½ teaspoon ground black pepper
¼ teaspoon celery seeds
One 4-pound rack baby back ribs, cut into 2 or 3 sections to fit in the cooker

Directions
1. **Preparing the Ingredients** Whisk the tomato paste, vinegar, paprika, coriander and fennel seeds, onion powder, thyme, allspice, salt, pepper, and celery seeds with ¾ cup water in an electric pressure cooker until the tomato paste dissolves. Add the ribs; toss to coat thoroughly and evenly in the sauce.

2. **High pressure for 32 minutes.** Lock the lid on the COSORI Pressure Cooker Multi-Cooker® and then cook for 32 minutes. To get 32-minutes cook time, press "Manual" button and use the TIME ADJUSTMENT button to adjust the cook time to 32 minutes.

3. **Pressure Release** Let the pressure to come down naturally for at least 15 minutes, then quick release any pressure left in the pot.

4. **Finish the dish** Unlock and open the pot. Transfer the rib rack sections to a large rimmed baking sheet. Set the electric one to its browning function. Bring the sauce to a simmer. Cook, stirring occasionally, until the sauce has thickened, 3 to 5 minutes. Position the oven rack 4 to 6 inches from the broiler; heat the broiler. Brush a light coating of the sauce onto the ribs, then broil until glazed and hot, 6 to 8 minutes, turning once. Slice the racks between the bones to make individual ribs. Serve with the extra sauce on the side.

Sausage And Peppers

PREP: 5 MINUTES • PRESSURE: 10 MINUTES • TOTAL: 15 MINUTES • PRESSURE LEVEL: HIGH • RELEASE: QUICK
SERVES 6

Ingredients

  2 tablespoons olive oil
  2½ pounds sweet Italian sausages in their casings
  4 large red bell peppers, stemmed, seeded, and cut into strips
  1 medium red onion, halved and thinly sliced
  2 medium garlic cloves, slivered
  1 cup red (sweet) vermouth
  2 tablespoons balsamic vinegar
  ¼ teaspoon grated nutmeg

Directions

1. **Preparing the Ingredients.** Heat the oil in a COSORI Pressure Cooker Multi-Cooker®, turned to the browning function. Prick the sausages with a fork, add them to the pot, and brown on all sides, about 6 minutes. Transfer to a large bowl.

   Add the peppers and onion; cook, stirring almost constantly, just until the pepper strips glisten, about 2 minutes. Add the garlic, cook a few seconds, then stir in the vermouth, vinegar, and nutmeg. Nestle the sausages into the mixture.

2. **High pressure for 10 minutes**. Lock the lid on the COSORI Pressure Cooker Multi-Cooker® and Cook for 10 minutes. To get 10-minutes cook time, press the "Steam Potatoes" button.

3. **Pressure Release** Use the quick-release method to bring the pot's pressure back to normal.

4. Unlock and open the pot. Stir well before serving.

Spicy Sausage And Chard Pasta Sauce

PREP: 5 MINUTES • PRESSURE: 6 MINUTES • TOTAL: 11 MINUTES • PRESSURE LEVEL: HIGH • RELEASE: QUICK

SERVES 6

Ingredients

2 tablespoons olive oil

1 medium red onion, chopped

Up to 3 small hot chiles, such as cherry peppers or Anaheim chiles, stemmed, seeded, and chopped

1 tablespoon minced garlic

1 pound mild Italian pork sausage meat, any casings removed

½ cup dry red wine, such as Syrah

½ cup canned tomato paste

¼ cup chicken broth

1 tablespoon dried basil

2 teaspoons dried oregano

4 cups stemmed and chopped Swiss chard

Directions

1. **Preparing the Ingredients**. Heat the oil in a COSORI Pressure Cooker Multi-Cooker®, turned to the browning function.

   Add the onion and cook, stirring often, until softened, about 4 minutes. Add the chiles and garlic; cook until aromatic, stirring all the while, about 1 minute.

   Crumble in the sausage meat, breaking up any clumps with a wooden spoon. Stir until it loses its raw color. Stir in the wine, tomato paste, broth, basil, and oregano until the tomato paste dissolves. Add the chard and stir well.

2. **High pressure for 6 minutes.** Lock the lid onto the cooker, set the machine's timer to cook at high pressure for 6 minutes. To get 6-minutes cook time, press the "Manual" button and use the TIME ADJUSTMENT button to adjust the cook time to 6 minutes.

3. **Pressure Release.** Use the quick-release method to drop the pressure back to normal.

4. Unlock and open the pot. Stir well before serving.

Ground Beef Stew

PREP: 5 MINUTES • PRESSURE: 5 MINUTES • PRESSURE LEVEL: HIGH • RELEASE: QUICK
SERVES 4

Ingredients

1 tablespoon olive oil

1½ pounds lean ground beef (about 93% lean)

1 large yellow onion, chopped

1 large sweet potato (about 1 pound), peeled and shredded through the large holes of a box grater

1 teaspoon ground cinnamon

1 teaspoon ground cumin

½ teaspoon dried sage

½ teaspoon dried oregano

½ teaspoon salt

½ teaspoon ground black pepper

2 tablespoons yellow cornmeal

2 tablespoons honey

2½ cups beef broth

Directions

1. **Preparing the Ingredients.** Heat the oil in the COSORI Pressure Cooker Multi-Cooker® turned to the "Browning" function. Crumble in the ground beef; cook, stirring occasionally, until it loses its raw color and browns a bit, about 5 minutes. Add the onion; cook, stirring often, until softened, about 3 minutes.

   Stir in the sweet potato, cinnamon, cumin, sage, oregano, salt, and pepper. Cook for 1 minute, stirring constantly. Stir in the cornmeal and honey; cook for 1 minute, stirring often, to dissolve the cornmeal. Stir in the broth.

2. **High pressure for 5 minutes**. Lock the lid on the COSORI Pressure Cooker Multi-Cooker® and then cook for 5 minutes. To get 5-minutes cook time, press "Manual" button and use the TIME ADJUSTMENT button to adjust the cook time to 5 minutes.

3. **Pressure Release** Use the quick-release method to drop the pot's pressure to normal.

4. Unlock and open the lid. Stir well and set aside, loosely covered, for 5 minutes before serving.

Lamb with Mexican Sauce

PREP: 10 MINUTES • PRESSURE: 45 MINUTES • TOTAL: 55 MINUTES • PRESSURE LEVEL: HIGH • RELEASE: NORMAL

SERVES 3-4

Ingredients

3 lamb shoulder
1 Spanish onion
3 garlic cloves, minced
1 19 oz. can Old El Paso Enchilada sauce
2 Tbsp. oil
Salt to taste
Cilantro, chopped without the stems
Corn tortillas (3-4 per person)
Limes cut into 8ths
Black beans or refried beans
Chipotle style rice

Directions

1. **Preparing the Ingredients** Marinate lamb overnight in Old El Paso Enchilada sauce (mild, medium or hot).
   Turn on the COSORI Pressure Cooker Multi-Cooker® to "sauté" mode.
   Add oil. Put in the onions and cook until soft, add garlic and cook for 1 minute.
   Add the lamb and marinade wait until boil.
2. **High pressure for 45 minutes**. Lock the lid on the COSORI Pressure Cooker Multi-Cooker® and then cook for 45 minutes. To get 45-minutes cook time, press Meat/Stew button and use the ADJUST button to adjust the cook time to 45 minutes.
3. **Pressure Release** Let the pressure to come down naturally for at least 15 minutes, then quick release any pressure left in the pot.
4. **Finish the dish**. Cut the limes, heat the beans, put the hot rice into a serving bowl.
   Set the Lamb aside. Ladle generous amount of sauce over it.
   Heat up 3-4 corn tortillas.
   Put the lamb mixture onto a soft warm corn tortilla, sprinkle on cilantro, then squeeze on lime juice.
5. Serve and Enjoy!

Pulled BBQ Beef Sandwiches

PREP: 10 MINUTES • PRESSURE: 35 MINUTES • TOTAL: 45 MINUTES • PRESSURE LEVEL: HIGH • RELEASE: NORMAL

SERVES 2-4

Ingredients

2 pounds – Beef of choice

2 cps – Water

4 cps – Finely shredded Cabbage (the secret ingredient and you'll never know it's in there.)

1/2 cup – Of your favorite BBQ Sauce

1 cup – Ketchup

1/3 cup – Worcestershire Sauce

1 tblsp – Horse Radish

1 tblsp – mustard

Directions

1. **Preparing the Ingredients.** Add and stir in ingredients to your COSORI Pressure Cooker Multi-Cooker®.
2. **High pressure for 35 minutes.** Lock the lid on the COSORI Pressure Cooker Multi-Cooker ® and then cook for 35 minutes. To get 35-minutes cook time, press "Manual" button.
3. **Pressure Release.** Use natural release method.
4. **Finish the dish** Set the beef aside. Set the COSORI Pressure Cooker Multi-Cooker® to a "Sauté" mode, Sauté the sauce until it reaches the desired consistency.
5. Serve and Enjoy.

One Pot Chinese Beef Stew

PREP: 12 MINUTES • PRESSURE: 30 MINUTES • TOTAL: 42 MINUTES • PRESSURE LEVEL: HIGH • RELEASE: NORMAL

SERVES 4-6

Ingredients

    1-2 Tsps. oil
    2 medium onions sliced
    ½ tsp sugar
    2 tsps. Rice wine or sherry
    1 Tb soy sauce
    1 kg beef round, cubed into one inch pieces
    2 tsps. cornstarch
    Pinch of smoked Paprika
    1-2 tsp garlic powder
    Salt and pepper
    ½ cup broth, preferably beef
    1 Tb Worcestershire sauce
    1 can of mushrooms
    1-2 tsps. Fresh ginger chopped finely
    1-2 tsps. Cornstarch slurry if needed

Directions

1. **Preparing the Ingredients.** Place sugar, rice wine and soy sauce into Power Pressure Cooker® using the "Sauté" mode fry for 30 seconds.
   Add beef broth and Worcestershire sauce, stir and close the lid.
2. **High pressure for 30 minutes.** Lock the lid on the COSORI Pressure Cooker Multi-Cooker® and then cook for 30 minutes. To get 30-minutes cook time, press Meat/Stew button and use the ADJUST button to adjust the cook time to 30 minutes.
   Leave on keep warm for 3 minutes.
3. **Pressure Release.** Release pressure using the Natural Release Method.
4. **Finish the dish** When meat is done, add chopped ginger, mushrooms (optional) and more salt and pepper (if needed).
   Sauté for another minute.
   Add cornstarch slurry to thicken to desired taste (if needed).
5. Serve with rice and stir fried greens or fresh cut veggies.
   Enjoy!

Glorious Beef Stew

PREP: 25 MINUTES • PRESSURE: 60 MINUTES • TOTAL: 70 MINUTES • PRESSURE LEVEL: HIGH • RELEASE: NORMAL

SERVES 4-6

Ingredients

2 pounds beef stew meat

2 packets McCormick Stew Seasoning (or stew seasoning of your choice for 2 pounds meat)

4 cups water

5 scrubbed medium-sized potatoes chopped

1 cup carrots chopped

1 onion chopped

4 stalks celery

1 cup raw green beans

Directions

1. **Preparing the Ingredients** Add the beef Stew meet, McCormick Stew Seasoning Packets, and the water to the COSORI Pressure Cooker Multi-Cooker ®
2. **High pressure for 45 minutes**. Lock the lid on the COSORI Pressure Cooker Multi-Cooker® and then cook for 45 minutes. To get 45-minutes cook time, press Meat/Stew button and use the ADJUST button to adjust the cook time to 45 minutes.
3. **Pressure Release.** Release the pressure using Natural Release.
   Remove the lid and stir.
4. Add vegetables below the maximum fill line, put lid back on.
5. **High pressure for 15 minutes**. Lock the lid on the COSORI Pressure Cooker Multi-Cooker® and cook for 15 minutes. To get 15-minutes cook time, press "Poultry" Button, and then "Adjust" Button.
6. **Pressure Release**. Use Natural Release Method.
7. Serve and enjoy.

Lamb And Eggplant Pasta Casserole

PREP: 10 MINUTES • PRESSURE: 8 MINUTES • TOTAL: 18 MINUTES • PRESSURE LEVEL: HIGH • RELEASE: QUICK

SERVES 4

Ingredients

2 tablespoons olive oil

1 medium red onion, chopped

1 tablespoon minced garlic

1½ pounds lean ground lamb

One small eggplant (about ¾ pound), stemmed and diced

¾ cup dry red wine, such as Syrah

2¼ cups chicken broth

½ cup canned tomato paste

1 teaspoon ground cinnamon

½ tablespoon dried oregano

½ teaspoon dried dill

½ teaspoon salt

½ teaspoon ground black pepper

8 ounces dried spiral-shaped pasta, such as rotini

Directions

1. **Preparing the Ingredients.** Heat the oil in the COSORI Pressure Cooker Multi-Cooker® turned to the "Browning" function. Add the onion and cook, stirring often, until softened, about 4 minutes. Add the garlic and cook until aromatic, less than 1 minute. Crumble in the ground lamb; cook, stirring occasionally, until it has lost its raw color, about 5 minutes. Add the eggplant and cook for 1 minute, stirring often, to soften a bit. Pour in the red wine and scrape up any browned bits in the pot as it comes to a simmer. Stir in the broth, tomato paste, cinnamon, oregano, dill, salt, and pepper until everything is coated in the tomato sauce. Stir in the pasta until coated.
2. **High pressure for 8 minutes**. Lock the lid on the COSORI Pressure Cooker Multi-Cooker® and then cook for 8 minutes. To get 8-minutes cook time, press "Manual" button and use the TIME ADJUSTMENT button to adjust the cook time to 8 minutes.
3. **Pressure Release.** Use the quick-release method.
4. Unlock and open the pot. Stir well before serving.

Lamb Shanks Provençal

PREP: 10 MINUTES • PRESSURE: 40 MINUTES • TOTAL: 50 MINUTES • PRESSURE LEVEL: HIGH • RELEASE: NATURAL

SERVES 6

Ingredients

2 large (12-ounce) lamb shanks
1 teaspoon kosher salt, plus additional for seasoning
Freshly ground black pepper
1 tablespoon olive oil
1 cup sliced onion
2 garlic cloves, finely minced
2 medium plum tomatoes, coarsely chopped, or ½ cup diced canned tomatoes, drained
½ cup dry white wine or dry white vermouth
1 cup Chicken Stock or low-sodium broth
1 bay leaf
1 lemon, sliced very thin
⅓ cup pitted Kalamata olives
2 tablespoons coarsely chopped fresh parsley

Directions

1. **Preparing the Ingredients.** Sprinkle the lamb shanks with 1 teaspoon of kosher salt and several grinds of pepper. The longer ahead of the cooking time you can do this, the better. Cover and let sit for 20 minutes to 2 hours at room temperature or refrigerate for up to 24 hours.

   Heat the vegetable oil in the COSORI Pressure Cooker Multi-Cooker® using the "Sauté" function, until the oil is shimmering and flows like water. Add the lamb shanks, and brown on all sides, about 6 minutes total. Remove them to a plate. Add the onion and garlic, and sprinkle with a pinch or two of kosher salt. Cook, stirring, for about 3 minutes, or until the onions just begin to brown. Add the tomatoes, and cook until most of their liquid evaporates.

   Add the white wine, and stir, scraping up the browned bits from the bottom of the cooker.

   Cook for 2 to 3 minutes, or until the wine reduces by about half; then add the Chicken Stock and bay leaf. Return the lamb shanks to the cooker, and place the lemon slices over them.

2. **High pressure for 40 minutes**. Lock the lid on the COSORI Pressure Cooker Multi-Cooker® and then cook for 40 minutes. To get 40-minutes cook time, press "Manual" button.

3. **Pressure Release.** After cooking, use the natural method to release pressure.

4. **Finish the dish.** Unlock and remove the lid. Transfer the lamb to a cutting board or plate, and tent it with aluminum foil. Strain the sauce into a fat separator, and let it rest until the fat rises to the surface.

If you don't have a fat separator, let the sauce sit for a few minutes, then spoon or blot off any excess fat from the top and discard. Pour the defatted sauce back into the cooker along with the strained vegetables. If you want a thicker sauce, simmer the liquid for about 5 minutes, or until it reaches the desired consistency.

Stir in the olives and parsley. Place the shanks in shallow bowls, pour the sauce and vegetables over the lamb, and serve.

Lamb shanks benefit from salting in advance, which makes them much more flavorful and helps them brown beautifully. If you have the time, salt them up to 24 hours in advance. Place them on a tray and refrigerate, covered loosely with foil.

Lamb Shanks With Pancetta

PREP: 15 MINUTES • PRESSURE: 60 MINUTES • TOTAL: 75 MINUTES • PRESSURE LEVEL: HIGH • RELEASE: NATURAL

SERVES 4

Ingredients

- 2 tablespoons olive oil
- One 6-ounce pancetta chunk, chopped
- Four 12-ounce lamb shanks
- 1 small yellow onion, chopped
- One 28-ounce can diced tomatoes, drained (about 3½ cups)
- 1 ounce dried mushrooms, preferably porcini, crumbled
- 3 tablespoons packed celery leaves, minced
- 2 tablespoons minced chives
- 2 cups dry, light white wine, such as Sauvignon Blanc
- 2 tablespoons all-purpose flour
- ½ teaspoon ground black pepper

Directions

1. **Preparing the Ingredients.** Heat the oil in the COSORI Pressure Cooker Multi-Cooker ®, turned to the "Browning" function. Add the pancetta and brown well, about 6 minutes, stirring often. Use a slotted spoon to transfer the pancetta to a large bowl. Add two of the shanks to the cooker; brown on all sides, turning occasionally, about 8 minutes. Transfer them to the bowl and repeat with the remaining shanks.

   Add the onion to the pot; cook, stirring often, until softened, about 4 minutes. Stir in the tomatoes, dried mushroom crumbles, celery leaves, and chives. Cook until bubbling, about minutes, stirring often.

   Whisk the wine, flour, and pepper in a medium bowl until the flour dissolves; stir this mixture into the sauce in the pot. Cook until thickened and bubbling, about 1 minute. Return the shanks, pancetta, and their juices to the cooker.

2. **High pressure for 60 minutes.** Close the lid and the pressure valve and then cook for 60 minutes. To get 60-minutes cook time, press "Manual" button and use the TIME ADJUSTMENT button to adjust the cook time to 60 minutes.

   Turn off the COSORI Pressure Cooker Multi-Cooker ® or unplug it so it doesn't jump to its keep-warm setting.

3. **Pressure Release.** Let its pressure return to normal naturally, 20 to 30 minutes.

4. **Finish the dish.** Unlock and open the cooker. Transfer a shank to each serving bowl. Skim any surface fat from the sauce with a flatware spoon. Ladle the sauce and vegetables over the lamb shanks.

Enchilada-Braised Chicken Breasts

PREP: 5 MINUTES • PRESSURE: 15 MINUTES • TOTAL: 19 MINUTES • PRESSURE LEVEL: HIGH • RELEASE: QUICK
SERVES 4

Ingredients

- 1 teaspoon packed dark brown sugar
- 1 teaspoon ground cumin
- 1 teaspoon smoked paprika
- ½ teaspoon salt
- ½ teaspoon ground black pepper
- ½ teaspoon onion powder
- ¼ teaspoon garlic powder
- Four 6- to 8-ounce boneless skinless chicken breasts
- 2 tablespoons olive oil
- One 8-ounce can tomato sauce (1 cup)
- ½ cup light-colored beer, preferably a Pilsner or an IPA
- 2 tablespoons chili powder
- 2 tablespoons fresh lime juice

Directions

1. **Preparing the Ingredients.** Mix the brown sugar, cumin, smoked paprika, salt, pepper, onion powder, and garlic powder in a medium bowl. Massage the spice rub onto the chicken breasts.

   Heat the oil in the COSORI Pressure Cooker Multi-Cooker® using the "Sauté" function. Set the breasts in the cooker and brown well, turning once, about 6 minutes.

   Mix the tomato sauce, beer, chili powder, and lime juice in the bowl the spices were in; pour the sauce over the breasts.

2. **High pressure for 15 minutes** Close the lid and Cook for 15 minutes. To get 15-minutes cook time, press the "Poultry" Button and the adjust button.

3. **Pressure Release** Use the quick-release method to bring the pot's pressure back to normal.

4. Unlock and open the cooker. Serve the chicken with the sauce ladled on top.

Honey-Chipotle Chicken Wings

PREP: 5 MINUTES • PRESSURE: 10 MINUTES • TOTAL: 15 MINUTES • PRESSURE LEVEL: HIGH • RELEASE: QUICK
SERVES 2

Ingredients

1 cup water, for steaming

3 tablespoons Mexican hot sauce (such as Valentina brand)

2 tablespoons honey

1 teaspoon minced canned chipotle in adobo sauce

Directions

1. **Preparing the Ingredients.** If using whole wings, cut off the tips and discard. Cut the wings at the joint into two pieces each, the "drumette" and the "flat."
   Add the water and insert the steamer basket or trivet. Place the wings on the steamer insert.
2. **High pressure for 10 minutes.** Close the lid and the pressure valve and then cook for 10 minutes. To get 10-minutes cook time, press "Manual" button and the Cook Time Selector.
3. **Pressure Release** Use the quick-release method.
4. **Finish the dish.** While the wings are cooking, make the sauce. In a large bowl, whisk together the hot sauce, honey, and minced chipotle. Preheat the broiler, and place an oven rack in the top or second position.
   Unlock and remove the lid. Using tongs, carefully transfer the wing segments to the bowl with the sauce. Toss gently to coat. Transfer the wing segments to a baking rack placed over a sheet pan, or to a baking sheet lined with nonstick aluminum foil.
   Place the baking sheet under the broiler for 4 to 5 minutes, or until the wings start to brown, and serve.

PER SERVING: CALORIES: 434; FAT: 27G; SODIUM: 1,152MG; CARBOHYDRATES: 19G; FIBER: 1G; PROTEIN: 31G

One Pot Pressure Cooker Chicken And Rice

PREP: 10 MINUTES • PRESSURE: 10 MINUTES • TOTAL: 20 MINUTES • PRESSURE LEVEL: HIGH • RELEASE: NATURAL

SERVES 2-4

Ingredients

6 dried shiitake mushrooms, marinated
6 - 8 chicken drumsticks, marinated
2 rice measuring cups (360 ml) Jasmine rice, rinse
1 teaspoon Salt
1½ cup (375 ml) water
1 tablespoon ginger, shredded
Green onions for garnish
Marinade:
1 tablespoon light soy sauce
1 teaspoon dark soy sauce
½ teaspoon sugar
½ teaspoon corn starch
1 teaspoon Shaoxing rice wine
A dash of white pepper powder
1 tablespoon ginger, shredded
1 teaspoon five spice powder

Directions

1. **Preparing the Ingredients.** Place the dried shiitake mushrooms in a small bowl. Rehydrate them with cold water for 20 minutes.

   Chop the drumsticks into 2 pieces. Then, marinate the chicken and mushrooms with the marinade sauce for 20 minutes.

   Rinse rice under cold water by gently scrubbing the rice with your fingertips in a circling motion. Pour out the milky water, and continue to rinse until the water is clear. Then, drain the water.

   Add the rice, 1 teaspoon of salt, and marinated chicken and mushrooms, and 1½ cup of water in the COSORI Pressure Cooker Multi-Cooker®.

2. **High pressure for 10 minutes.** Lock the lid on the COSORI Pressure Cooker Multi-Cooker® and then cook for 10 minutes. To get 10-minutes cook time, press "Manual" button.

3. **Pressure Release** Let the pressure to come down naturally for at least 15 minutes, then quick release any pressure left in the pot.

4. Serve immediately.

Lemon and Olive Ligurian Chicken

PREP: 10 MINUTES • PRESSURE: 15 MINUTES • TOTAL: 25 MINUTES • PRESSURE LEVEL: HIGH • RELEASE: NORMAL

SERVES 6-8

Ingredients

2 garlic cloves, chopped

3 sprigs of Fresh Rosemary (two for chopping, one for garnish)

2 sprigs of Fresh Sage

½ bunch of Parsley Leaves and stems

3 lemons, juiced (about ¾ cup or 180ml)

4 tablespoons extra virgin olive oil

1 teaspoon sea salt

¼ teaspoon pepper

1 whole chicken, cut into parts or package of bone-in chicken pieces, skin removed (or not) ½ cup (125ml) dry white wine

3.5oz (100g) Black Gourmet Salt-Cured Olives (Taggiesche , French, or Kalamata)

1 fresh lemon, for garnish (optional)

Directions

1. **Preparing the Ingredients.** Prepare the marinade by finely chopping together the garlic, rosemary, sage, and parsley. Place them in a container and add the lemon juice, olive oil, salt and pepper. Mix well and set aside.

   Remove the skin from the chicken (save it for a chicken stock).

   In the preheated COSORI Pressure Cooker Multi-Cooker®, with the lid off, add a swirl of olive oil and brown the chicken pieces on all sides for about 5 minutes.

   De-glaze cooker with the white wine until it has almost all evaporated (about 3 minutes).

   Add the chicken pieces back in - this time being careful with the order. Put all dark-meat (wings, legs, thighs) first, and then the chicken breasts on top so that they do not touch the bottom of the COSORI Pressure Cooker Multi-Cooker ®.

   Pour the remaining marinade on top. Don't worry if this does not seem like enough liquid, the chicken will also release its juices into the cooker, too.

2. **High pressure for 10 minutes.** Lock the lid on the COSORI Pressure Cooker Multi-Cooker® and then cook for 10 minutes. To get 10-minutes cook time, press "Manual" button.

3. **Pressure Release.** When time is up, open the cooker by releasing the pressure using the Quick-Release Method.

4. **Finish the dish.** Take the chicken pieces out of the cooker and place on a serving platter tightly covered with foil.

   Reduce the cooking liquid in the COSORI Pressure Cooker Multi-Cooker ®, with the lid off to ¼ of its amount, or until it becomes thick and syrupy.

Put all of the chicken pieces back into the COSORI Pressure Cooker Multi-Cooker® to warm up. Mix and spoon the thick glaze onto the chicken pieces and simmer it in the glaze for a few minutes before serving.

Sprinkle with fresh rosemary, olives and lemon slices. When serving, caution your guests that the olives still have their pits!

Per Serving Calories: 204.8; Fat: 12.2g; Carbohydrates: 3.1g; Sugar: 0.7g; Fiber: 0.3g; Protein: 17.8g; Sodium: 449.5mg; Cholesterol: 61.6mg

Chicken with Artichoke Hearts and Mushrooms

PREP: 5 MINUTES • PRESSURE: 12 MINUTES • TOTAL: 17 MINUTES • PRESSURE LEVEL: HIGH • RELEASE: NATURAL
SERVES 4-6

Ingredients

½ teaspoon kosher salt
2 (8-ounce) or 4 (4-ounce) bone-in, skin-on chicken thighs
1 tablespoon olive oil
¼ cup sliced onion
4 ounces white button or cremini mushrooms, trimmed and quartered
½ cup dry white wine
1 bay leaf
¼ teaspoon dried thyme
½ cup frozen artichoke hearts, thawed
⅓ cup low-sodium chicken broth
Freshly ground black pepper

Directions

1. **Preparing the Ingredients.** Using ½ teaspoon of kosher salt, sprinkle the chicken thighs on both sides.
   In the COSORI Pressure Cooker Multi-Cooker® set to "browning", heat the olive oil until it shimmers and flows like water. Add the chicken thighs, skin-side down, and cook, undisturbed, for about 6 minutes, or until the skin is dark golden brown and most of the fat under the skin has rendered. Turn the thighs to the other side, and cook for about 3 minutes more, or until that side is light golden brown. Remove the thighs. Carefully pour off almost all the fat, leaving just enough (about 1 tablespoon) to cover the bottom of the COSORI Pressure Cooker Multi-Cooker ® with a thick coat. Add the onion and mushrooms, and cook for about 5 minutes, or until softened. Add the white wine, and cook for 3 to 5 minutes, or until reduced by half. Add the bay leaf, thyme, artichokes, and chicken broth, and bring to a simmer. Return the chicken to the pot, skin-side up.
2. **High pressure for 12 minutes.** Lock the lid on the COSORI Pressure Cooker Multi-Cooker® and then cook for 12 minutes. To get 12-minutes cook time, press "Manual" Button and use the TIME ADJUSTMENT button to adjust the cook time to 12 minutes.
3. **Pressure Release.** After cooking, use the natural method to release pressure.
4. **Finish the dish.** Unlock and remove the lid. Remove the chicken thighs from the pan, and set aside. Remove the bay leaf. Strain the sauce into a fat separator, and let it rest until the fat rises to the surface. If you don't have a fat separator, let the sauce sit for a few minutes; then spoon or blot off any excess fat from the top and discard. Pour the defatted sauce back into the cooker, and add the chicken thighs and the solids from the sauce. If you prefer a thicker sauce, turn the COSORI Pressure Cooker Multi-Cooker ® to the "Sauté" function, and simmer the sauce for several minutes until it's reduced to the consistency you like.

Adjust the seasoning, adding more salt if necessary and several grinds of pepper, and serve.

PER SERVING: CALORIES: 612; FAT: 38G; SODIUM: 1,269MG; CARBOHYDRATES: 19G; FIBER: 3G; PROTEIN: 37G

Chicken Breasts With White Wine And Orange Juice

PREP: 5 MINUTES • PRESSURE: 18 MINUTES • TOTAL: 25 MINUTES • PRESSURE LEVEL: HIGH • RELEASE: QUICK

SERVES 4

Ingredients

3 tablespoons unsalted butter
Four 12-ounce bone-in, skin-on chicken breasts
½ teaspoon salt
½ teaspoon ground black pepper
½ cup fresh orange juice
½ cup dry but light white wine, such as Sauvignon Blanc
One 4-inch fresh rosemary sprig
1 tablespoon honey
½ tablespoon potato starch or cornstarch

Directions

1. **Preparing the Ingredients.** Melt the butter in a COSORI Pressure Cooker Multi-Cooker®, turned to the browning function. Season the chicken with the salt and pepper, then add two breasts skin side down to the cooker. Brown well, turning once, about 5 minutes; transfer to a large bowl. Brown the remaining breasts, and leave them in the cooker.

   Return the first two breasts to the cooker, arranging them so that all are skin up but overlapping only as necessary, thinner parts over thick. Pour the orange juice and wine over the chicken. Tuck in the rosemary and drizzle everything with honey.

2. **High pressure for 18 minutes.** Lock the lid on the COSORI Pressure Cooker Multi-Cooker® and then cook for 18 minutes. To get 18-minutes cook time, press "Manual" button and use the TIME ADJUSTMENT button to adjust the cook time to 18 minutes.

3. **Pressure Release** Use the quick-release method to bring the pot's pressure back to normal.

4. **Finish the dish.** Unlock and open the pot. Discard the rosemary sprig. Use kitchen tongs to transfer the chicken breasts to individual serving plates or a serving platter.

   Dissolve the potato starch or cornstarch with ½ tablespoon water in a small bowl. Turn the electric cooker to its browning function, bring the sauce to a simmer. Add this slurry and cook, stirring all the time, until thickened, about 20 seconds. Ladle the sauce over the chicken to serve.

Honey Soy Chicken Wings

PREP: 10 MINUTES • PRESSURE: 20 MINUTES • TOTAL: 30 MINUTES • PRESSURE LEVEL: HIGH • RELEASE: NATURAL

SERVES 2-4

Ingredients

1 ½ pound chicken wings
4 cloves garlic, roughly minced
½ large shallot or 1 small shallot, roughly minced
1 – 2 star anise
1 tablespoon ginger, sliced
1 tablespoon honey
½ cup warm water
1 tablespoon peanut oil
1 ½ tablespoon cornstarch
Chicken Wing Marinade
2 tablespoons light soy sauce
1 tablespoon dark soy sauce
1 tablespoon Shaoxing wine
1 teaspoon sugar
¼ teaspoon salt

Directions

1. **Preparing the Ingredients**. Marinate the chicken wings with the Chicken Wing Marinade for 20 minutes.

   Heat the COSORI Pressure Cooker Multi-Cooker® using the "Sauté" function.

   Add 1 tablespoon of peanut oil into the pot. Add the marinated chicken wings into the pot. Then, brown the chicken wings for roughly 30 seconds on each side. Flip a few times as you brown them as the soy sauce and sugar can be burnt easily. Remove and set aside.

   Add the minced shallot, star anise and sliced ginger, then stir for roughly a minute. Add the minced garlic and stir until fragrant (roughly 30 seconds).

   Mix 1 tablespoon of honey with ½ cup of warm water, then add it into the pot and deglaze the bottom of the pot with a wooden spoon.

   Place all the chicken wings with all the meat juice and the leftover chicken wing marinade into the pot.

2. **High pressure for 5 minutes**. Lock the lid on the COSORI Pressure Cooker Multi-Cooker® and then cook for 5 minutes. To get 5-minutes cook time, press "Manual" Button and use the TIME ADJUSTMENT button to adjust the cook time to 5 minutes.

3. **Pressure Release** Let the pressure to come down naturally for at least 10 minutes, then quick release any pressure left in the pot.

4. **Finish the dish.** Open the lid carefully and taste one of the honey soy chicken wings and the honey soy sauce. Season with more salt or honey if desired.

Remove all the chicken wings from the pot and set aside. Turn the COSORI Pressure Cooker Multi-Cooker® to its browning function. Mix 1 ½ tablespoon of cornstarch with 1 tablespoon of cold running tap water. Keep mixing and add it into the honey soy sauce one third at a time until desired thickness.

Turn off the heat and add the chicken wings back into the pot. Coat well with the honey soy sauce and serve immediately!

Thai Green Chicken Curry

PREP: 15 MINUTES • PRESSURE: 10 MINUTES • TOTAL: 25 MINUTES • PRESSURE LEVEL: HIGH • RELEASE: QUICK
SERVES 6-8

Ingredients

- 1 tablespoon vegetable oil
- 1 medium onion, peeled, and sliced thin
- 3 cloves garlic, crushed
- 1/2 inch piece of ginger, peeled and crushed
- Cream from the top of a (13.5 ounce) can coconut milk
- 4 tablespoons green curry paste (a whole 4 oz. can)
- 3 pounds boneless skinless chicken thighs, cut into 1/2 inch by 2 inch lengths
- 1 teaspoon Diamond Crystal kosher salt or 3/4 teaspoon fine sea salt
- The rest of the (13.5 ounce) can coconut milk
- 1 cup chicken stock or water
- 1 tablespoon fish sauce (plus more to taste)
- 1 tablespoon soy sauce (plus more to taste)
- 1 tablespoon brown sugar (plus more to taste)
- Juice from 1 lime
- 12 ounces green beans, trimmed and cut into 2 inch pieces
- Minced cilantro
- Minced basil (preferably Thai basil)
- Lime wedges
- Jasmine rice

Directions

1. **Preparing the Ingredients.** Heat the vegetable oil in the COSORI Pressure Cooker Multi-Cooker® until shimmering, use Sauté mode. Stir in the onion, garlic, and ginger, and Sauté until the onion starts to soften, about 3 minutes.

   Fry the curry paste: Scoop the cream from the top of the can of coconut milk and add it to the pot, then stir in the curry paste. Cook, stirring often, until the curry paste darkens, about 5 minutes.

   Sprinkle the chicken with the kosher salt. Add the chicken to the pot, and stir to coat with curry paste. Stir in the rest of the can of coconut milk, chicken stock, fish sauce, soy sauce, and brown sugar.

2. **High pressure for 10 minutes.** Lock the lid on the COSORI Pressure Cooker Multi-Cooker® and then cook for 10 minutes. To get 10-minutes cook time, press "Manual" Button and use the TIME ADJUSMENT button to adjust the cook time to 10 minutes.

3. **Pressure Release** Use the quick-release method to bring the pot's pressure back to normal.

4. **Finish the dish.** Finish the curry: Remove the lid from the COSORI Pressure Cooker Multi-Cooker®, then set Sauté mode. Stir in the lime juice and the green beans, and simmer the curry until the green beans are crisp-tender, about 4 minutes. Taste the

curry for seasoning, adding more soy sauce (to add salt) or brown sugar (to add sweet) as needed. Ladle the curry into bowls, sprinkle with minced cilantro and basil, and serve with Jasmine rice.

5. Enjoy!

Penne with Chicken

PREP: 5 MINUTES • PRESSURE: 5 MINUTES • TOTAL: 10 MINUTES • PRESSURE LEVEL: HIGH • RELEASE: QUICK

SERVES 4

Ingredients

1 tablespoon all-purpose flour

1 teaspoon kosher salt, divided

⅛ teaspoon granulated garlic or garlic powder

½ teaspoon dried Italian herbs, divided (or ¼ teaspoon dried oregano and ¼ teaspoon dried basil)

⅛ teaspoon freshly ground black pepper

3 (4-ounce) boneless, skinless chicken thighs

1 tablespoon olive oil

1 cup thinly sliced onion

1 small green bell pepper, seeded and cut into 1-inch chunks (about 1½ cups)

3 garlic cloves, minced or pressed (about 1 tablespoon)

½ cup dry white or red wine

1½ cups Quick Marinara Sauce or plain tomato sauce

2 tablespoons minced sun-dried tomatoes (optional)

1¾ cups water

½ pound penne or similar pasta shape

3 cups arugula or baby spinach

Parmigiano-Reggiano or a similar cheese, for garnish

Directions

1. **Preparing the Ingredients.** In a small bowl or jar with a shaker top, mix together the flour, ½ teaspoon of kosher salt, the granulated garlic, ¼ teaspoon of Italian herbs, and the pepper. Sprinkle the flour mixture over both sides of the chicken thighs, coating as evenly as possible.

   Set COSORI Pressure Cooker Multi-Cooker® to "brown", heat the olive oil until it shimmers and flows like water. Add the chicken thighs, and cook for 5 minutes, or until golden brown. Turn the thighs over, and cook the other side for 5 minutes more, or until that side is also golden brown. Remove the thighs to a rack or cutting board, and cool for 3 minutes.

   With the COSORI Pressure Cooker Multi-Cooker® on "brown," add the onion, green bell pepper, and garlic. Cook for about 3 minutes, stirring, until the onions just start to brown. Pour in the wine, and scrape the bottom of the pan to release the browned bits, cooking until the wine is almost completely evaporated. Add the Quick Marinara Sauce, the remaining ½ teaspoon of kosher salt, the sun-dried tomatoes (if using), the remaining ¼ teaspoon of Italian herbs, the water, the chicken, and the penne.

2. **High pressure for 5 minutes.** Lock the lid on the COSORI Pressure Cooker Multi-Cooker® and then cook for 5 minutes. To get 5-minutes cook time, press "Manual" button and use the TIME ADJUSTMENT button to adjust the cook time to 5 minutes.

3. **Pressure Release** Use the quick-release method.
4. **Finish the dish.** Unlock and remove the lid. The penne should be almost done, and the sauce will be a little thin. Add the arugula, and stir. With the COSORI Pressure Cooker Multi-Cooker® set to "Sauté", cook for 3 to 4 minutes, or until the pasta is done to your liking, the arugula is wilted, and the sauce has thickened. Serve topped with grated Parmigiano-Reggiano.

   PER SERVING: CALORIES: 510; FAT: 14G; SODIUM: 1,040MG; CARBOHYDRATES: 63G; FIBER: 6G; PROTEIN: 35G

Main Dishes – Turkey

Easy Turkey Breast
Turkey Meatballs In A Creamy Tomato Sauce
Speedy Turkey Chili
Speedy Texas Trail Chili
Spicy Turkey Chili
Turkey Breast
Paleo Turkey and Gluten Free Gravy
Turkey Sloppy Joes

Easy Turkey Breast

PREP: 10 MINUTES • PRESSURE: 60 MINUTES • TOTAL: 70 MINUTES • PRESSURE LEVEL: HIGH • RELEASE: NATURAL

SERVES 4

Ingredients

1 frozen turkey breast with frozen gravy packet
1 whole onion

Directions

1. **Preparing the Ingredients**. Place frozen turkey breast, rozen gravy packet and whole onion in the COSORI Pressure Cooker Multi-Cooker®.
2. **High pressure for 30 minutes.** Lock the lid on the COSORI Pressure Cooker Multi-Cooker® and then cook for 30 minutes. To get 30-minutes cook time, press "Manual" button and use the TIME ADJUSTMENT button to adjust the cook time to 30 minutes.
3. **Pressure Release.** Use natural-release method.
   Remove lid, turn turkey breast over
4. **High pressure for 30 minutes.** Replace lid on the COSORI Pressure Cooker Multi-Cooker® and then cook for 30 minutes. To get 30-minutes cook time, press "Manual" button
   and use the TIME ADJUSTMENT button to adjust the cook time to 30 minutes.
5. **Pressure Release.** Use natural-release method, again.
6. **Finish the dish.** Remove mesh. Remove turkey and slice. Places slices and turkey gravy into serving dish.
7. Enjoy!

Turkey Meatballs In A Creamy Tomato Sauce

PREP: 5 MINUTES • PRESSURE: 10 MINUTES • TOTAL: 15 MINUTES • PRESSURE LEVEL: HIGH • RELEASE: QUICK

SERVES 4

Ingredients

1 pound ground turkey

1 large egg, at room temperature and beaten in a small bowl

½ cup plain dried breadcrumbs

¼ cup finely grated Parmesan cheese (about ½ ounce)

½ teaspoon dried oregano

½ teaspoon dried rosemary

½ teaspoon ground black pepper

½ teaspoon salt

2 tablespoons unsalted butter

1 medium yellow onion, chopped

2 medium celery stalks, thinly sliced

One 28-ounce can whole tomatoes, drained and roughly chopped (about 3½ cups)

½ cup chicken broth

1 tablespoon packed fresh oregano leaves, minced

¼ cup heavy cream

¼ teaspoon grated nutmeg

Directions

1. **Preparing the Ingredients.** Mix the ground turkey, egg, breadcrumbs, cheese, oregano, rosemary, pepper, and ¼ teaspoon salt in a large bowl until well combined. Form the mixture into 12 balls.

   Melt the butter in the COSORI Pressure Cooker Multi-Cooker® turned to the browning function. Add the onion and celery; cook, stirring often, until the onion turns translucent, about 3 minutes.

   Stir in the tomatoes, broth, oregano, and the remaining ¼ teaspoon salt. Drop the meatballs into the sauce.

2. **High pressure for 10 minutes**. Lock the lid on the COSORI Pressure Cooker Multi-Cooker® and then cook for 10 minutes. To get 10-minutes cook time, press "Manual" Button and use the TIME ADJUSTMENT button to adjust the cook time to 10 minutes.

3. **Pressure Release** Use the quick-release method to drop the pot's pressure to normal.

4. **Finish the dish** Unlock and open the cooker. Turn the COSORI Pressure Cooker Multi-Cooker® to its browning function. Stir in the cream and nutmeg; simmer, stirring all the while, for 1 minute to reduce the cream a little and blend the flavors.

Speedy Turkey Chili

PREP: 30 MINUTES • PRESSURE: 10 MINUTES • TOTAL: 40 MINUTES • PRESSURE LEVEL: LOW • RELEASE: NATURAL

SERVES 4

Ingredients

1 lb. 85% lean ground turkey

4 to 5 ounces water

15 ounces chick peas (or your favorite white bean), previously cooked in your COSORI Pressure Cooker Multi-Cooker ®!

1 yellow bell pepper diced (you can add another yellow bell pepper)

I medium onion diced

2 – 3 cloves garlic peeled and not chopped 2.5 TBSP chili powder

1.5 tsp cumin

1/8 tsp cayenne

2 cans original rotel

One 5.5 ounce can V8

12 ounces water with vegetable stock or 12-ounces vegetable stock

Directions

1. **Preparing the Ingredients** Add lean ground turkey and water into COSORI Pressure Cooker Multi-Cooker®.
2. **High pressure for 5 minutes.** Lock the lid on the COSORI Pressure Cooker Multi-Cooker® and then cook for 5 minutes. To get 5-minutes cook time, press "Manual" Button and use the TIME ADJUSTMENT button to adjust the cook time to 5 minutes.
3. **Pressure Release.** Use natural-release method for 10 minutes, then quick-release. Open the COSORI Pressure Cooker Multi-Cooker ® and break up the ground turkey add the remaining ingredients.
4. **High pressure for 5 minutes.** Lock the lid on the COSORI Pressure Cooker Multi-Cooker® and then cook for 5 minutes. To get 5-minutes cook time, press "Manual" Button and use the TIME ADJUSTMENT button to adjust the cook time to 5 minutes.
5. **Pressure Release.** Use natural-release method for 10 minutes, then quick-release.
6. Stir, Serve and enjoy.

Speedy Texas Trail Chili

PREP: 15 MINUTES • PRESSURE: 5 MINUTES • TOTAL: 20 MINUTES • PRESSURE LEVEL: HIGH • RELEASE: QUICK
SERVES 8

Ingredients

2 Tablespoons canola oil
1 large onion, peeled, chopped
1-1/2 pounds ground beef, turkey or chicken
2 Cups favorite Bloody Mary mix (spicy preferred)
2 Cans (14 ounces each) diced tomatoes with green chilies (or 28-ounce can diced tomatoes with juice)
2 Cans (14 ounces each) kidney beans, drained and rinsed well
4 Tablespoons (or more if you like) favorite chili powder, divided
1-1/2 Cups Water
Corn chips
Shredded cheese
Sliced green onions
Sour cream

Directions

1. **Preparing the Ingredients.** In the COSORI Pressure Cooker Multi-Cooker ® pressure cooker pot, heat the oil. Use "Sauté" Function. Add the onion and Sauté about 8 minutes until it becomes lightly golden brown. Add the meat and cook until it browns, breaking it up as it cooks.
   When meat is done, remove turkey thighs to cutting board and cover loosely with foil.
   Stir in the Bloody Mary mix and heat, stirring and scraping up any browned bits on the bottom of the pan.
   Add the tomatoes, beans and 2 Tablespoons of the chili powder. Stir well. Bring to just a boil; add the water.
2. **High pressure for 5 minutes.** Lock the lid on the COSORI Pressure Cooker Multi-Cooker® and then cook for 5 minutes. To get 5-minutes cook time, press "Manual" Button and use the TIME ADJUSTMENT button to adjust the cook time to 5 minutes.
3. **Pressure Release.** Use the Quick-Release method.
4. **Finish the dish** Just before serving stir in Tablespoons of the chili powder then let it stand 5 minutes.
   Ladle into bowls and garnish as desired.
5. Serve and Enjoy!

Spicy Turkey Chili

PREP: 10 MINUTES • PRESSURE: 45 MINUTES • TOTAL: 55 MINUTES • PRESSURE LEVEL: HIGH • RELEASE: NATURAL

SERVES 4

Ingredients

- 1 tablespoon olive oil
- 1 medium yellow onion, diced
- 2 green bell peppers, seeded and diced
- 2 fresh cayenne peppers, chopped (seeds included)
- 4 cloves garlic, chopped
- 1 teaspoon ground cumin
- ½ teaspoon dried oregano leaves
- 1 pound ground turkey
- ¼ cup your favorite hot sauce
- 1 (15-ounce) can fire-roasted diced tomatoes
- 1 (15-ounce) can kidney beans, including their liquid
- 1 cup grated Monterey Jack cheese
- ¼ cup chopped cilantro

Directions

1. **Preparing the Ingredients.** Set the COSORI Pressure Cooker Multi-Cooker ® to its "Sauté" setting and add the oil. Add the onions, peppers, and garlic, and sauté until the onions soften and begin to brown, about 10 minutes.

   Add the cumin and oregano and sauté two more minutes, until aromatic.

   Add the ground turkey, breaking it up with a spoon or spatula. Sauté until opaque and cooked through, about 5 minutes.

   Add the hot sauce, canned tomatoes and kidney beans and stir to combine.
2. **High pressure for 45 minutes**. Lock the lid on the COSORI Pressure Cooker Multi-Cooker® and then cook for 45 minutes. To get 45-minutes cook time, press Meat/Stew button, and use the ADJUST button to adjust the cook time to 45 minutes.
3. **Pressure Release** Use natural-release method.
4. **Finish the dish** Top with grated cheese and cilantro, and serve with rice or cornbread, if desired.
5. Enjoy!

Turkey Breast

PREP: 20 MINUTES • PRESSURE: 45 MINUTES • TOTAL: 65 MINUTES • PRESSURE LEVEL: HIGH • RELEASE: NATURAL & QUICK
SERVES 4

Ingredients

6.5 lb. bone-in, skin-on turkey breast
Salt and pepper, to taste
1 (14 oz.) can turkey or chicken broth
1 large onion, quartered
1 stock celery, cut in large pieces
1 sprig thyme
3 tablespoons cornstarch
3 tablespoons cold water

Directions

1. **Preparing the Ingredients.** Season turkey breast liberally with salt and pepper. Put trivet in the bottom. Add chicken broth, onion, celery and thyme. Add the turkey to the cooking pot breast side up.
2. **High pressure for 45 minutes.** Lock the lid on the COSORI Pressure Cooker Multi-Cooker® and then cook for 45 minutes. To get 45-minutes cook time, press Meat/Stew button and use the ADJUST button to adjust the cook time to 45 minutes.
3. **Pressure Release.** Use a natural pressure release for 10 minutes, then do a quick pressure release. Check if the turkey is done. If it isn't, lock the lid in place and cook it for a few more minutes.
4. **Finish the dish.** Carefully remove turkey and place on large plate. Cover with foil. Strain and skim the fat off the broth. Whisk together corn starch and cold water; add to broth in cooking pot. Select Sauté and stir until broth thickens. Add salt and pepper to taste. Slice the turkey and serve immediately.
5. Enjoy!

Paleo Turkey and Gluten Free Gravy

PREP: 10 MINUTES • PRESSURE: 35 MINUTES • TOTAL: 45 MINUTES • PRESSURE LEVEL: HIGH • RELEASE: QUICK
SERVES 6

Ingredients

1 4-5 pound bone-in, skin-on turkey breast
Salt
Black pepper (omit for AIP)
2 tablespoons ghee or butter (use coconut oil for AIP)
1 medium onion, cut into medium dice
1 large carrot, cut into medium dice
1 celery rib, cut into medium dice
1 garlic clove, peeled and smashed
2 teaspoons dried sage
¼ cup dry white wine
1½ cups bone broth (preferably from chicken or turkey bones)
1 bay leaf
1 tablespoon tapioca starch (optional)

Directions

1. **Preparing the Ingredients.** Set the "Sauté" function.
   Pat turkey breast dry and generously season with salt and pepper. Melt cooking fat in the COSORI Pressure Cooker Multi-Cooker ®.
   Brown turkey breast, skin side down, about 5 minutes, and transfer to a plate, leaving fat in pot.
   Add onion, carrot, and celery to pot and cook until softened, about 5 minutes. Stir in garlic and sage and cook until fragrant, about 30 seconds.
   Pour in wine and cook until slightly reduced, about 3 minutes. Stir in broth and bay leaf. Using wooden spoon, scrape up all browned bits stuck on bottom of pot.
   Place turkey skin side up in post with any accumulated juices.
2. **High pressure for 35 minutes**. Lock the lid on the COSORI Pressure Cooker Multi-Cooker® and then cook for 35 minutes. To get 35-minutes cook time, press "Manual" button, and use the TIME ADJUSTMENT button to adjust the cook time to 35
3. **Pressure Release** Use quick-release method and carefully remove lid.
4. **Finish the dish** Transfer turkey breast to carving board or plate and tent loosely with foil, allowing it to rest while you prepare the gravy.
   Use an immersion blender or carefully transfer cooking liquid and vegetables to blender and puree until smooth. Return to heat and cook until thickened and reduced to about 2 cups. Adjust seasoning to taste.
   Slice turkey breast and serve with hot gravy.
5. Enjoy!

Turkey Sloppy Joes

PREP: 5 MINUTES • PRESSURE: 30 MINUTES • TOTAL: 35 MINUTES • PRESSURE LEVEL: HIGH • RELEASE: NATURAL
SERVES 2

Ingredients

- 1 tablespoon olive oil
- ¼ cup chopped onion
- ¼ small red or green bell pepper, chopped (about 2 tablespoons)
- 1 garlic clove, minced or pressed
- Kosher salt
- ⅔ cup tomato sauce
- ¼ cup beer or water
- 1 tablespoon cider or wine vinegar, plus additional as needed
- 1 tablespoon packed brown sugar
- 2 teaspoons ancho or New Mexico chili powder
- 1 teaspoon Dijon mustard
- ½ teaspoon Worcestershire sauce
- 1 large or 2 small turkey thighs (1½ pounds total), skin removed
- 2 hamburger buns

Directions

1. **Preparing the Ingredients.** Set the COSORI Pressure Cooker Multi-Cooker ® to "brown", heat the olive oil until it shimmers and flows like water. Add the onion, bell pepper, and garlic, and sprinkle with a pinch or two of kosher salt. Cook for about 5 minutes, stirring, until the onions just begin to brown. Add the tomato sauce, beer, cider vinegar, brown sugar, chili powder, mustard, and Worcestershire sauce. Bring to a simmer. Stir to make sure the brown sugar is dissolved. Place the turkey thigh in the cooker.
2. **High pressure for 30 minutes.** Lock the lid on the COSORI Pressure Cooker Multi-Cooker® and then cook for 30 minutes. To get 30-minutes cook time, press Meat / Stew button and use the TIME ADJUSTMENT button to adjust the cook time to 30 minutes.
3. **Pressure Release.** Use the natural-release.
   Unlock and remove the lid; transfer the turkey to a plate or cutting board to cool.
4. **Finish the dish.** Set the COSORI Pressure Cooker Multi-Cooker® to "brown,", and simmer the sauce for about 5 minutes, or until it's the consistency of a thick tomato sauce. Skim any visible fat from the surface and discard. Taste and adjust the seasoning.
5. Serve on the hamburger buns.
6. Enjoy!

PER SERVING: CALORIES: 532; FAT: 20G; SODIUM: 1,253MG; CARBOHYDRATES: 36G; FIBER: 5G; PROTEIN: 53G

Hominy, Peppers, And Pork Stew

PREP: 5 MINUTES • PRESSURE: 12 MINUTES • TOTAL: 17 MINUTES • PRESSURE LEVEL: HIGH • RELEASE: QUICK
SERVES 4

Ingredients

2 tablespoons olive oil
1 large yellow or white onion, chopped
1 large green bell pepper, stemmed, cored, and cut into ¼-inch-thick strips
1 large red bell pepper, stemmed, cored, and cut into ¼-inch-thick strips
2 teaspoons minced garlic
2 teaspoons minced, seeded fresh jalapeño chile
2 teaspoons dried oregano
2½ cups canned hominy, drained and rinsed
One 14-ounce can diced tomatoes, drained (about 1¾ cups)
1 cup chicken broth
1 pound boneless center-cut pork loin chops, cut into ¼-inch-thick strips

Directions

1. **Preparing the Ingredients.** Heat the oil in a COSORI Pressure Cooker Multi-Cooker®, turned to the browning function. Add the onion and both bell peppers; cook, stirring often, until the onion softens, about 4 minutes.
   Add the garlic, jalapeño, and oregano; stir well until aromatic, less than 20 seconds. Add the hominy, tomatoes, broth, and pork; stir over the heat for 1 minute.
2. **High pressure for 12 minutes.** Lock the lid on the COSORI Pressure Cooker Multi-Cooker® and then cook for 12 minutes. To get 12-minutes cook time, press "Manual" button and use the TIME ADJUSTMENT button to adjust the cook time to 12 minutes.
3. **Pressure Release** Use the quick-release method to bring the pot's pressure back to normal.
4. Unlock and open the cooker. Stir well before serving.

Pork Shoulder Chops With Soy Sauce, Maple Syrup, And Carrots

PREP: 5 MINUTES • PRESSURE: 40 MINUTES • TOTAL: 45 MINUTES • PRESSURE LEVEL: HIGH • RELEASE: NATURAL
SERVES 6

Ingredients

1 tablespoon bacon fat
3 pounds bone-in pork shoulder chops, each ½ to ¾ inch thick
6 medium carrots
3 medium garlic cloves
⅓ cup soy sauce
⅓ cup maple syrup
⅓ cup chicken broth
½ teaspoon ground black pepper

Directions

1. **Preparing the Ingredients.** Melt the bacon fat in a COSORI Pressure Cooker Multi-Cooker®, turned to the browning function. Add about half the chops and brown well, turning once, about 5 minutes. Transfer these to a large bowl and brown the remaining chops.
   Stir the carrots and garlic into the pot; cook for 1 minute, stirring constantly. Pour in the soy sauce, maple syrup, and broth, stirring to dissolve the maple syrup and to get up any browned bits on the bottom of the pot. Stir in the pepper. Return the shoulder chops and their juices to the pot. Stir to coat them in the sauce.

2. **High pressure for 40 minutes**. Lock the lid on the COSORI Pressure Cooker Multi-Cooker® and then cook for 40 minutes. To get 40-minutes cook time, press "Manual" button and use the TIME ADJUSTMENT button to adjust the cook time to 40 minutes.

3. **Pressure Release** Let the pressure to come down naturally for at least 14 to 16 minutes, then quick release any pressure left in the pot.

4. **Finish the dish** Unlock and open the pot. Transfer the chops, carrots, and garlic cloves to a large serving bowl. Skim the fat off the sauce and ladle it over the servings.

Pulled Pork

PREP: 5 MINUTES • PRESSURE: 80 MINUTES • TOTAL: 85 MINUTES • PRESSURE LEVEL: HIGH • RELEASE: NATURAL
SERVES 8

Ingredients

2 tablespoons smoked paprika
2 tablespoons packed dark brown sugar
1 tablespoon ground cumin
2 teaspoons ground black pepper
½ tablespoon dry mustard
1 teaspoon ground coriander
1 teaspoon dried thyme
1 teaspoon onion powder
1 teaspoon salt
½ teaspoon garlic powder
½ teaspoon ground cloves
½ teaspoon ground cinnamon
One 4- to 4½-pound bone-in skinless pork shoulder, preferably pork butt
Up to 1½ cups light-colored beer, preferably a pale ale or amber lager

Directions

1. **Preparing the Ingredients.** Mix the smoked paprika, brown sugar, cumin, pepper, mustard, coriander, thyme, onion powder, salt, garlic powder, cloves, and cinnamon in a small bowl. Massage the mixture all over the pork.
   Set the pork in the COSORI Pressure Cooker Multi-Cooker®. Pour 1 cup beer into the electric cooker without knocking the spices off the meat.
2. **High pressure for 80 minutes**. Lock the lid on the COSORI Pressure Cooker Multi-Cooker® and then cook for 80 minutes. To get 80-minutes cook time, press "Manual" button and use the TIME ADJUSTMENT button to adjust the cook time to 80 minutes.
3. **Pressure Release** Let its pressure fall to normal naturally, 25 to 35 minutes.
4. **Finish the dish** Transfer the meat to a large cutting board. Let stand for 5 minutes. Use a spoon to skim as much fat off the sauce in the pot as possible.
   Set the "browning "function. Bring the sauce to a simmer, stirring occasionally; continue boiling the sauce, stirring often, until reduced by half, 7 to 10 minutes.
   Use two forks to shred the meat off the bones; discard the bones and any attached cartilage. Pull any large chunks of meat apart with the forks and stir the meat back into the simmering sauce to reheat.
5. Serve and Enjoy!

Pork Carnitas

PREP: 15 MINUTES • PRESSURE: 50 MINUTES • TOTAL: 65 MINUTES • PRESSURE LEVEL: HIGH • RELEASE: NATURAL
SERVES 11

Ingredients

2 1/2 pounds trimmed, boneless pork shoulder blade roast
2 teaspoons kosher salt
black pepper, to taste
6 cloves garlic, cut into sliver
1 1/2 teaspoons cumin
1/2 teaspoon sazon
1/4 teaspoon dry oregano
3/4 cup reduced sodium chicken broth
2-3 chipotle peppers in adobo sauce (to taste)
2 bay leaves
1/4 teaspoon dry adobo seasoning
1/2 teaspoon garlic powder

Directions

1. **Preparing the Ingredients.** Season pork with salt and pepper. Bring the cooker to high pressure by pressing the Brown/ Sauté button, and brown pork on all sides on high heat for about 5 minutes. Remove from heat and allow to cool.
   Using a sharp knife, insert blade into pork about 1-inch deep, and insert the garlic slivers, you'll want to do this all over. Season pork with cumin, sazon, oregano, adobo and garlic powder all over.
   Pour chicken broth, add chipotle peppers and stir, add bay leaves and place pork in the COSORI Pressure Cooker Multi-Cooker®.
6. **High pressure for 50 minutes.** Lock the lid on the COSORI Pressure Cooker Multi-Cooker® and then cook for 50 minutes. To get 50-minutes cook time, press "Manual" button and use the TIME ADJUSTMENT button to adjust the cook time to 50 minutes.
2. **Pressure Release**. Use natural-release method.
3. Serve and Enjoy!

Per Serving Calories: 160; Fat: 7g; Sat Fat: 3g; Carb: 1g; Fiber: 0g; Protein: 20g; Sugar: 0g; Sodium: 397 mg; Cholesterol: 69mg

Easy Pork Chops

PREP: 15 MINUTES • PRESSURE: 5 MINUTES • TOTAL: 20 MINUTES • PRESSURE LEVEL: LOW • RELEASE: NATURAL
SERVES 6

Ingredients
3-4 Pork Chops – ½ to ¾ inch thick
One egg – beaten
Flour
Salt and Pepper
Bread Crumbs
Onions – chopped – as much as you like – ½ cup maybe
2 – 4 Garlic cloves – squashed and chopped
Butter- 1 tbsp.
Oil 1-2 tbsp. or orange/ginger coconut oil

Directions

1. **Preparing the Ingredients.** Turn on the COSORI Pressure Cooker Multi-Cooker® to the Sauté setting, then wait for it to boil. Heat the oil and butter to very hot.

   Make sure your pork chops are at room temperature. Dredge them in flour, dip into beaten egg, dredge them in bread crumbs. Brown them lots on both sides in the hot COSORI Pressure Cooker Multi-Cooker®. When well browned on both sides, remove and put on plate.

   Throw in the onions, swish them around for a minute until softer looking, then throw in the garlic and swish around.

   Leave the onions, garlic and drippings in the pot. Add about two to three tablespoons of water. Put steamer in pot, place browned pork chops on steamer above the water and drippings.

2. **High pressure for 5 minutes.** Lock the lid on the COSORI Pressure Cooker Multi-Cooker® and then cook for 5 minutes. To get 5-minutes cook time, "Manual" button and use the TIME ADJUSTMENT button to adjust the cook time to 5 minutes.

3. **Pressure Release** Let the pressure to come down naturally for at least 15 minutes, then quick release any pressure left in the pot.

4. **Finish the dish.** Remove from the pot. Perfect, juicy pork chops you may use the 'juice' in the pot to pour over the pork chops or you can add a little polenta (or flour) and water, Sauté and make it like a gravy.

5. Serve and enjoy!

Pork Chops With Applesauce

PREP: 10 MINUTES • PRESSURE: 10 MINUTES • TOTAL: 20 MINUTES • PRESSURE LEVEL: HIGH • RELEASE: NATURAL

SERVES 2-4

Ingredients

2 – 4 pork loin chops (we used center cut, bone-on)
1 tablespoon grapeseed oil or olive oil
1 small onion, sliced
3 cloves garlic, roughly minced
2 gala apples, thinly sliced
2 pieces whole cloves (optional)
1 teaspoon cinnamon powder
1 tablespoon honey
½ cup unsalted homemade chicken stock or water
2 tablespoons light soy sauce
1 tablespoon butter
Kosher salt and ground black pepper to taste
1 ½ tablespoon cornstarch mixed with 2 tablespoons water (optional)

Directions

1. **Preparing the Ingredients.** Make a few small cut around the sides of the pork chops so they will stay flat and brown evenly.
   Season the pork chops with generous amount of kosher salt and ground black pepper.
   Heat up your COSORI Pressure Cooker Multi-Cooker®. Add grapeseed oil into the pot. Add the seasoned pork chops into the pot, then let it brown for roughly 2 – 3 minutes on each side. Remove and set aside.
   Add in the sliced onions and stir. Add a pinch of kosher salt and ground black pepper to season if you like. Cook the onions for roughly 1 minute until soften. Then, add garlic and stir for 30 seconds until fragrance.
   Add in the thinly sliced gala apples, whole cloves (optional) and cinnamon powder, then give it a quick stir. Add the honey and partially deglaze the bottom of the pot with a wooden spoon.
   Add chicken stock and light soy sauce, then fully deglaze the bottom of the pot with a wooden spoon. Taste the seasoning and add more salt and pepper if desired.
   Place the pork chops back with all the meat juice into the pot.
2. **High pressure for 10 minutes.** Lock the lid on the COSORI Pressure Cooker Multi-Cooker® and then cook for 10 minutes. To get 10-minutes cook time, press "Manual" Button and use the TIME ADJUSTMENT button to adjust the cook time to 10 minutes.
3. **Pressure Release**. Let it fully natural release (roughly 10 minutes). Open the lid carefully.
4. **Finish the dish.** Remove the pork chops and set aside. Turn the pressure cooker to the Sauté setting. Remove the cloves and taste the seasoning one more time. Add more salt and pepper if desired. Add butter and stir until it has fully dissolved into the sauce.

Mix the cornstarch with water and mix it into the applesauce one third at a time until desired thickness.
Drizzle the applesauce over the pork chops and serve immediately with side dishes!

Spare Ribs With Wine

PREP: 5 MINUTES • PRESSURE: 15 MINUTES • TOTAL: 20 MINUTES • PRESSURE LEVEL: HIGH • RELEASE: NATURAL
SERVES 2-4

Ingredients

1 pound pork spare ribs, cut into pieces
1 tablespoon oil
1 tablespoon corn starch
1 – 2 teaspoon water
Green onions as garnish
1 teaspoon fish sauce (optional)
Black Bean Marinade:
1 tablespoon black bean sauce
1 tablespoon light soy sauce
1 tablespoon Shaoxing wine
1 tablespoon ginger, grated
3 cloves garlic, minced
1 teaspoon sesame oil
1 teaspoon sugar
A pinch of white pepper

Directions

1. **Preparing the Ingredients.** Marinate the pork spare ribs with Black Bean Marinade in an oven-safe bowl. Then, sit it in the fridge for 25 minutes.

   First, mix 1 tablespoon of oil into the marinated spare ribs. Then, add 1 tablespoon of corn starch and mix well. Finally, add 1 – 2 teaspoon of water into the spare ribs and mix well.

   Add 1 cup of water into the COSORI Pressure Cooker Multi-Cooker®. Place steam rack in the COSORI Pressure Cooker Multi-Cooker®. Then, put the bowl of spare ribs on the rack.

2. **High pressure for 15 minutes.** Lock the lid on the COSORI Pressure Cooker Multi-Cooker® and then cook for 15 minutes. To get 15-minutes cook time, press "Poultry" Button and then adjust button.

3. **Pressure Release** Let the pressure to come down naturally for at least 15 minutes, then quick release any pressure left in the pot.

4. **Finish the dish.** Taste and add one teaspoon of fish sauce and green onions as garish if you like.

5. Serve immediately.

Pork Loin With Apples

PREP: 5 MINUTES • PRESSURE: 30 MINUTES • TOTAL: 35 MINUTES • PRESSURE LEVEL: HIGH • RELEASE: QUICK

SERVES 6-8

Ingredients

2 tablespoons unsalted butter

One 3-pound boneless pork loin roast

1 large red onion, halved and thinly sliced

2 medium tart green apples, such as Granny Smith, peeled, cored, and thinly sliced

4 fresh thyme sprigs

2 bay leaves

½ cup moderately sweet white wine, such as Riesling

¼ cup chicken broth

½ teaspoon salt

½ teaspoon ground black pepper

Directions

1. **Preparing the Ingredients.** Melt the butter in the COSORI Pressure Cooker Multi-Cooker®, set on the "Browning"function. Add the pork loin and brown it on all sides, turning occasionally, about 8 minutes in all. Transfer to a large plate.

   Add the onion to the pot; cook, stirring often, until softened, about 3 minutes. Stir in the apple, thyme, and bay leaves. Pour in the wine and scrape up any browned bits on the bottom of the pot.

   Pour in the broth; stir in the salt and pepper. Nestle the pork loin into this apple mixture; pour any juices from the plate into the pot.

2. **High pressure for 30 minutes.** Lock the lid on the COSORI Pressure Cooker Multi-Cooker® and then cook for 30 minutes. To get 30-minutes cook time, press "Manual" Button.

3. **Pressure Release** Use the quick-release method to bring the pot's pressure to normal.

4. **Finish the dish** Unlock and open the cooker. Discard the bay leaves. Transfer the pork to a cutting board; let stand for 5 minutes while you dish the sauce into serving bowls or onto a serving platter. Slice the loin into ½-inch-thick rounds and lay these over the sauce.

Pork Tenderloin And Coconut Rice

PREP: 5 MINUTES • PRESSURE: 15 MINUTES • TOTAL: 20 MINUTES • PRESSURE LEVEL: HIGH • RELEASE: QUICK

SERVES 4

Ingredients

2 tablespoons peanut oil
1 pound pork tenderloin, cut into 4 pieces
1 small leek, white and pale green parts only, halved lengthwise, washed and thinly sliced
One 4½-ounce can chopped mild green chiles (about ½ cup)
1 teaspoon dried thyme
1 teaspoon ground cumin
½ teaspoon ground coriander
¼ teaspoon salt
¼ teaspoon ground black pepper
One 15-ounce can black beans, drained and rinsed (about 1¾ cups)
1 cup chicken broth
1 cup regular or low-fat canned coconut milk
1 cup white long-grain rice, such as white basmati rice
2 tablespoons packed light brown sugar

Directions

1. **Preparing the Ingredients.** Heat the oil in the COSORI Pressure Cooker Multi-Cooker® turned to the "Browning"function. Add the pork tenderloin pieces; brown on all sides, turning occasionally, about 6 minutes. Transfer to a plate.
   Add the leek and chiles; cook, stirring often, until the leek softens, about 2 minutes. Stir in the thyme, cumin, coriander, salt, and pepper; cook until aromatic, less than half a minute. Stir in the beans, broth, coconut milk, rice, and brown sugar until the brown sugar dissolves.
   Nestle the pieces of pork in the sauce, submerging the meat and rice as much as possible in the liquid; pour any juices from the meat's plate into the cooker.
2. **High pressure for 15 minutes**. Lock the lid on the COSORI Pressure Cooker Multi-Cooker® and then cook for 15 minutes. To get 15-minutes cook time, press "Poultry" Button and then adjust button.
3. **Pressure Release** Use the quick-release method to bring the pot's pressure back to normal, but do not open the cooker.
   Set the pot aside for 10 minutes to steam the rice.
4. **Finish the dish** Unlock and open the cooker. Transfer the pork pieces to four serving plates; spoon the rice and beans around them.
5. Enjoy!

Pork Tenderloin with Braised Apples

PREP: 5 MINUTES • PRESSURE: 45 MINUTES • TOTAL: 50 MINUTES • PRESSURE LEVEL: HIGH • RELEASE: QUICK
SERVES 4

Ingredients.

For The Brine (optional)
½ cup Diamond Crystal kosher salt, or ¼ cup fine table salt
¼ cup granulated sugar
2 cups very hot tap water
2 cups ice water

For The Pork And Apples
1 (1-pound) pork tenderloin, trimmed of silver skin and halved crosswise
Kosher salt, for salting and seasoning
2 tablespoons unsalted butter
1 cup thinly sliced onion
1 medium Granny Smith apple, or other tart apple, peeled and cut into ¼-inch slices
¾ cup apple juice, cider, or hard cider
½ cup low-sodium chicken broth
2 tablespoons heavy (whipping) cream
1 teaspoon Dijon mustard, plus additional as needed

Directions

1. **Preparing the Ingredients**
   - To make the brine (if using)
   In a large stainless steel or glass bowl, dissolve the salt and sugar in the hot water; then stir in the ice water. Submerge the pork in the brine, and refrigerate for 2 to 3 hours. Drain and pat dry.
   - To make the pork and apples
   If you choose not to brine the pork, sprinkle it liberally with kosher salt.
   Set to Sauté/browning, heat the butter just until it stops foaming. Add the pork halves, browning on all sides, about 4 minutes total. Transfer to a plate or rack, and set aside. Add the onion slices to the cooker, and cook, stirring, for 2 to 3 minutes, or until they just start to brown. Add the apple slices, and cook for 1 minute. Add the apple juice, and scrape the browned bits from the bottom of the pot. Bring to a simmer, and cook for 2 to 3 minutes, or until the juice has reduced by about one-third. Add the chicken broth, and return the pork tenderloin to the cooker, placing the pieces on top of the apples and onions.

2. **High pressure for 45 minutes.** Lock the lid on the COSORI Pressure Cooker Multi-Cooker® and then cook for 45 minutes. To get 45-minutes cook time, press Meat/Stew button and use the ADJUST button to adjust the cook time to 45 minutes.

3. **Pressure Release.** Use the quick-release method.
4. **Finish the dish.** Unlock and remove the lid. Transfer the pork to a plate or rack, and tent it with aluminum foil while you finish the sauce.

   Turn the COSORI Pressure Cooker Multi-Cooker® to "brown," simmer for about 6 minutes, or until the liquid is reduced by about half. Stir in the heavy cream and mustard, and taste, adding kosher salt or more mustard as needed.

   Slice the pork into ¾-inch pieces, and place on a serving platter. Spoon the apples, onions, and sauce over the pork, and serve.

PER SERVING: CALORIES: 321; FAT: 13G; SODIUM: 754MG; CARBOHYDRATES: 21G; FIBER: 2G; PROTEIN: 32G

Cream of Asparagus Soup

PREP: 10 MINUTES • PRESSURE: HIGH • TIME UNDER PRESSURE: 5 MINUTES • RELEASE: NATURAL
SERVES: 4-6

Ingredients

- 1 tablespoon olive oil
- 3 green onions, sliced crosswise into ¼-inch pieces
- 1 pound asparagus, tough ends removed, cut into 1-inch pieces
- 4 cups salt-free Chicken Stock
- 1 tablespoon unsalted butter
- 1 tablespoon all-purpose flour
- 2 teaspoons salt
- 1 teaspoon ground white pepper, plus more as needed
- ½ cup heavy cream

Directions

1. **Preparing the Ingredients.** Heat the COSORI Pressure Cooker Multi-Cooker® using the "Sauté" function, add the oil, green onions, and a pinch of salt. Sauté the green onions for a few minutes, then add the asparagus and stock.
2. **High pressure for 5 minutes**. Lock the lid on the COSORI Pressure Cooker Multi-Cooker® and then cook for 5 minutes. To get 5-minutes cook time, press "Manual" Button and use the TIME ADJUSTMENT button to adjust the cook time to 5 minutes.
   Meanwhile, make a blond roux: In a small saucepan over low heat, mix together the butter and flour and cook, stirring constantly, until the butter has melted and the mixture foams and begins to turn golden beige. Remove from the heat.
3. **Pressure Release.** When the time is up, open the cooker with the Natural Release method.
4. **Finish the dish.** Add the roux, salt, and pepper to the soup and puree with an immersion blender until smooth. Taste and season with more pepper if you wish. Swirl in the cream just before serving.
5. Serve and Enjoy!

Chicken Soup

PREP: 10 MINUTES • PRESSURE: 35 MINUTES • TOTAL: 45 MINUTES • PRESSURE LEVEL: HIGH • RELEASE: NATURAL

SERVES 8

Ingredients

2 frozen boneless skinless chicken breasts

4 washed medium size diced potatoes (I did not peel you can if you want)

3 peeled carrots chopped into similar size as potatoes for even cooking time

1/2 large onion diced

4 cups of water and chicken concentrate/bullion of your choice to equal 32 ounces – or if you have it, use chicken stock

Salt and pepper to taste (flavors will intensify while under pressure)

Directions

1. **Preparing the Ingredients** Mix the broth, chicken, potatoes, onion, carrots, salt, and pepper in the COSORI Pressure Cooker Multi-Cooker®.
2. **High pressure for 35 minutes.** Lock the lid on the COSORI Pressure Cooker Multi-Cooker® and then cook for 35 minutes. To get 35-minutes cook time, press Soup button and use the TIME ADJUSTMENT button to adjust the cook time to 35 minutes.
3. **Pressure Release** Let the pressure to come down naturally for at least 15 minutes, then quick release any pressure left in the pot.
4. Open when all pressure is released stir and enjoy.

Chicken Stock

PREP: 10 MINUTES • PRESSURE: 60 MINUTES • TOTAL: 70 MINUTES • PRESSURE LEVEL: HIGH • RELEASE: NATURAL

SERVES 10 Cups

Ingredients

2 ½ pounds chicken carcasses
2 onions (keep the outer layers too), diced
2 celery stalks, diced
2 carrots, diced
2 bay leaves
4 garlic cloves, crushed
1 teaspoon whole peppercorn
10 cups water
Your favorite fresh herbs
1 tablespoon apple cider vinegar (optional)

Directions

1. **Preparing the Ingredients.** Optional step: Brown the chicken carcasses in your COSORI Pressure Cooker Multi-Cooker® with 1 tablespoon of oil. This will slightly elevate the flavors and result in a brown stock. Then, add water to deglaze the pot with 100 ml of water.
   Add all ingredients in the COSORI Pressure Cooker Multi-Cooker®
2. **High pressure for 60 minutes**. Lock the lid on the COSORI Pressure Cooker Multi-Cooker® and then cook for 60 minutes. To get 60-minutes cook time, press "Manual" Button and use the TIME ADJUSTMEN button to adjust the cook time to 60 minutes.
3. **Pressure Release.** When the time is up, open the cooker with the Natural Release method
4. **Finish the dish** Open the lid. Strain the stock through a colander discarding the solids, and set aside to cool. Let the stock sit in the fridge until the fat rises to the top and form a layer of gel. Then, skim off the fat on the surface.
   You can use the stock immediately, keep it in the fridge, or freeze it for future use.
   Storage: -Silicone Mold – We love freezing our chicken stock with this mold!! After they freeze in the mold, we pop them out and store them in Ziploc freezer bags. It's a great portion for many recipes, thaws quickly, and super convenient.

French Onion Soup

PREP: 5 MINUTES • PRESSURE: 35 MINUTES • TOTAL: 40 MINUTES • PRESSURE LEVEL: HIGH • RELEASE: NATURAL
SERVES 2-4

Ingredients

2 tablespoons unsalted butter, divided
4 cups thinly sliced white or yellow onions, divided
½ teaspoon kosher salt, plus additional for seasoning
¼ cup dry sherry
2 cups low-sodium chicken broth
½ cup Beef Stock, Mushroom Stock, or low-sodium broth
½ teaspoon Worcestershire sauce
¼ teaspoon dried thyme
1 teaspoon sherry vinegar or red wine vinegar, plus additional as needed
1 ounce Gruyère or other Swiss-style cheese, coarsely grated (about ⅓ cup)
2 thin slices French or Italian bread

Directions

1. **Preparing the Ingredients.** Set the COSORI Pressure Cooker Multi-Cooker® to "brown," heat 1 tablespoon of butter until it stops foaming, and then add 1 cup of onions. Sprinkle with a pinch or two of kosher salt, and stir to coat with the butter. Cook the onions in a single layer for about 4 minutes, or until browned. Resist the urge to stir them until you see them browning. Stir them to expose the other side to the heat, and cook for 4 minutes more. The onions should be quite browned but still slightly firm. Remove the onions from the pan, and set aside.
Pour the sherry into the pot, and stir to scrape up the browned bits from the bottom. When the sherry has mostly evaporated, add the remaining 1 tablespoon of butter, and let it melt. Stir in the remaining 3 cups of onions, and sprinkle with ½ teaspoon of kosher salt.
2. **High pressure for 25 minutes.** Lock the lid on the COSORI Pressure Cooker Multi-Cooker® and then cook for 25 minutes. To get 25-minutes cook time, press "Manual" Button and use the TIME ADJUSTMENT button to adjust the cook time to 25 minutes.
3. **Pressure Release** Use the quick-release method.
Unlock and remove the lid.
The onions should be pale and very soft, with a lot of liquid in the pot. Add the chicken broth, Beef Stock, Worcestershire sauce, and thyme.
4. **High pressure for 10 minutes**. Lock the lid on the COSORI Pressure Cooker Multi-Cooker® and then cook for 10 minutes. To get 10-minutes cook time, press "Manual" Button and use the TIME ADJUSTMENT button to adjust the cook time to 10 minutes.
5. **Pressure Release** Use the quick-release method.
6. **Finish the dish.** Unlock and remove the lid. Stir in the sherry vinegar, and taste. The soup should be balanced between the sweetness of the onions, the savory stock, and the acid from the vinegar. If it seems bland, add a pinch or two of kosher salt or a little more

vinegar. Stir in the reserved cup of onions, and keep warm while you prepare the cheese toasts.

7. Preheat the broiler. Reserve 2 tablespoons of the cheese, and sprinkle the remaining cheese evenly over the 2 bread slices. Place the bread slices on a sheet pan under the broiler for 2 to 3 minutes, or until the cheese melts.

   Place 1 tablespoon of the reserved cheese in each of 2 bowls. Ladle the soup into the bowls, float a toast slice on top of each, and serve.

PER SERVING: CALORIES: 366; FAT: 17G; SODIUM: 1,122MG; CARBOHYDRATES: 40G; FIBER: 6G; PROTEIN: 14G

Roasted Tomato Soup

PREP: 5 MINUTES • PRESSURE: HIGH • TIME UNDER PRESSURE: 10 MINUTES • RELEASE: QUICK
SERVES: 2

Ingredients.

3 tablespoons olive oil
½ cup sliced onion
Kosher salt
1 medium garlic clove, sliced or minced
¼ cup dry or medium-dry sherry
1 (14.5-ounce) can fire-roasted tomatoes
1 small roasted red bell pepper, cut into chunks (about ¼ cup)
¾ cup Chicken Stock or low-sodium broth
⅛ teaspoon ground cumin
⅛ teaspoon freshly ground black pepper
1 tablespoon heavy (whipping) cream (optional)

Directions

1. **Preparing the Ingredients.** Set the COSORI Pressure Cooker Multi-Cooker ® to brown, heat the olive oil until it shimmers and flows like water. Add the onions, and sprinkle with a pinch or two of kosher salt. Cook for about 5 minutes, stirring, until the onions just begin to brown. Add the garlic, and cook for 1 to 2 minutes more, or until fragrant.
Pour in the sherry, and simmer for 1 to 2 minutes, or until the sherry is reduced by half, scraping up any browned bits from the bottom of the pan. Add the tomatoes, roasted red bell pepper, and Chicken Stock to the COSORI Pressure Cooker Multi-Cooker®
2. **High pressure for 10 minutes**. Lock the lid on the COSORI Pressure Cooker Multi-Cooker® and then cook for 10 minutes. To get 10-minutes cook time, press "Manual" Button and use the TIME ADJUSTMENT button to adjust the cook time to 10 minutes.
3. **Pressure Release** Use the quick-release method.
4. **Finish the dish** For a smooth soup, blend using an immersion or standard blender. Add the cumin and pepper, and adjust the salt, if necessary. If you like a creamier soup, stir in the heavy cream.
 If using a standard blender, be careful. Steam can build up and blow the lid off if the soup is very hot. Hold the lid on with a towel, and blend in batches, if necessary; don't fill the jar more than halfway full.

> **PER SERVING:** CALORIES: 287; FAT: 24G; SODIUM: 641MG; CARBOHYDRATES: 16G; FIBER: 4G; PROTEIN: 4G

Butternut Squash Soup with Chicken Orzo

PREP: 5 MINUTES • PRESSURE: HIGH • TIME UNDER PRESSURE: 20 MINUTES • RELEASE: QUICK
SERVES: 6

Ingredients

1 ½ pounds of fresh baked butternut squash, peeled and cubed
1 tomato diced
3 tablespoons butter
1 onion, diced
1 garlic clove, minced
½ cup celery, diced
½ cup carrots, diced
2 cans chicken broth
2 tablespoon red pepper flakes
2 tablespoon dried parsley flakes
¼ teaspoon freshly ground black pepper
1 cup orzo, cooked
1 cup chicken breast, seasoned, cooked and diced

Directions

1. **Preparing the Ingredients.** Set the COSORI Pressure Cooker Multi-Cooker ® to brown, and melt butter to sauté the onion, garlic clove, celery and carrots.
   Then add the chicken broth, red pepper flakes, dried parsley flakes, black pepper, baked butternut squash and tomato diced to the COSORI Pressure Cooker Multi-Cooker®
2. **High pressure for 15 minutes.** Lock the lid on the COSORI Pressure Cooker Multi-Cooker® and then cook for 15 minutes. To get 15-minutes cook time, press "Poultry" Button.
3. **Pressure Release** Use the quick-release method.
   Blend/puree until mixture is smooth.
4. **High pressure for 5 minutes.** Then add it back to your COSORI Pressure Cooker Multi-Cooker® along with the chicken breast and orzo and cook for another 5 minutes. To get 5-minutes cook time, press "Manual" Button.
5. **Pressure Release** Use the quick-release method.
6. Serve with fresh dinner rolls and butter on the side

BEEF STOCK

Ingredients

2 lb (907 g) beef soup bones
3 large carrots
1 large onion, quartered, skin on
1 bay leaf
3 celery sticks
Handful fresh parsley
2 tsp (5 g) ground pepper
1 tsp (5 g) ground Himalayan salt
2 tbsp (19 g) garlic, minced
3 tbsp (45 ml) apple cider vinegar
Water

Directions

1. **Preparing the Ingredients.** Ideally, baking the bones at 375°F (190°C) for 30 minutes prior to pressure cooking them helps draw out the marrow, but if you only have access to your pressure cooker, it will still get the job done. To start the stock, place the bones, veggies and seasonings into the COSORI Pressure Cooker Multi-Cooker®. Pour in the apple cider vinegar and cover with water. The amount of water will vary based on the size and quantities of your vegetables. You can add in extra greens if you want.

2. **High pressure for 90 minutes.** Lock the lid on the COSORI Pressure Cooker Multi-Cooker® and then cook for 90 minutes. To get 90-minutes cook time, press "Manual" button and use the TIME ADJUSTMENT button to adjust the cook time to 90 minutes.

3. **Pressure Release** Once complete, quick-release the pressure valve, allowing the steam to escape.

Cream Of Sweet Potato Soup

PREP: 5 MINUTES • PRESSURE: HIGH • TIME UNDER PRESSURE: 15 MINUTES • RELEASE: QUICK
SERVES: 6

Ingredients

- 8 tablespoons (1 stick) unsalted butter, cut into small pieces
- 2 pounds sweet potatoes (about 2 large), peeled and cut into 2-inch pieces
- 1 teaspoon salt
- ½ teaspoon ground cinnamon
- ½ teaspoon ground ginger
- ¼ teaspoon baking soda
- 2½ cups chicken broth
- ½ cup heavy cream

Directions

1. **Preparing the Ingredients.** Melt the butter in a COSORI Pressure Cooker Multi-Cooker® turned to the browning function. Stir in the sweet potatoes, salt, cinnamon, ginger, and baking soda. Pour ½ cup water over everything.
2. **High pressure for 15 minutes.** Lock the lid on the COSORI Pressure Cooker Multi-Cooker® and then cook for 15 minutes. To get 15-minutes cook time, press "Poultry" Button and then adjust button.
3. **Pressure Release** Use the quick-release method to bring the pot's pressure back to normal.
4. **Finish the dish.** Unlock and open the pot. Stir in the broth and cream. Use an immersion blender to puree the soup in the pot; or ladle the soup in batches into a blender, remove the knob from the blender's lid, cover the hole with a clean kitchen towel, and blend until smooth.

Vegetable Stock

PREP: 5 MINUTES • PRESSURE: HIGH • TIME UNDER PRESSURE: 10 MINUTES • RELEASE: NATURAL
SERVES: 6

Ingredients

2 large unpeeled yellow onions, sliced lengthwise in half, root ends removed
2 medium carrots, snapped in half
2 celery stalks, snapped in half
2 medium tomatoes (fresh or canned)
1 bunch fresh flat-leaf parsley, tied with string (so it's easy to remove)
2 unpeeled garlic
1 tablespoon whole black peppercorns
2 bay leaves
Cold water, as needed

Directions

1. **Preparing the Ingredients.** Add the vegetables, herbs, and spices to the pressure cooker base. Pour in cold water to just cover these ingredients
2. **High pressure for 10 minutes**. Lock the lid on the COSORI Pressure Cooker Multi-Cooker® and then cook for 10 minutes. To get 10-minutes cook time, press "Manual" Button and use the TIME ADJUSTMENT button to adjust the cook time to 10 minutes.
3. **Pressure Release** Let the pressure to come down naturally for at least 20 to 30 minutes, then quick release any pressure left in the pot.
4. **Finish the dish.** Carefully strain the contents of the cooker into a stainless steel bowl and let cool to room temperature. Reserve the solids or discard them. Freeze the stock if not using in the next couple of days.

Chicken Noodle Soup

PREP: 5 MINUTES • PRESSURE: HIGH • TIME UNDER PRESSURE: 22 MINUTES • RELEASE: QUICK
SERVES: 6

Ingredients

Ingredients
2 tablespoons olive oil
2 large bone-in skinless chicken breasts (about 1 pound each)
6 cups chicken broth
1 medium red onion, halved
2 medium carrots
½ teaspoon salt
2 fresh thyme sprigs
2 fresh sage sprigs
2 medium garlic cloves, peeled
4 ounces wide egg noodles
1 tablespoon minced fresh dill fronds

Directions

1. **Preparing the Ingredients.** Heat the oil in the COSORI Pressure Cooker Multi-Cooker®, turned to the browning function. Add the chicken and brown well on both sides, about 4 minutes in all, turning once.
   Pour in the broth; add the onion, carrots, salt, thyme, sage, and garlic.
2. **High pressure for 18 minutes.** Lock the lid on the COSORI Pressure Cooker Multi-Cooker® and then cook for 18 minutes. To get 18-minutes cook time, press "Manual" Button and use the TIME ADJUSTMENT button to adjust the cook time to 18 minutes.
3. **Pressure Release** Use the quick-release method to return the pot's pressure to normal. Unlock and open the cooker. Transfer the chicken to a cutting board. Cool for a few minutes, then debone and chop the meat into bite-size bits; set aside.
   Discard the onion, carrots, thyme, sage, and garlic from the pot. Stir in the noodles and dill.
4. **High pressure for 4 minutes**. Lock the lid on the COSORI Pressure Cooker Multi-Cooker® and then cook for 4 minutes. To get 4-minutes cook time, press Steam button and use the TIME ADJUSTMENT button to adjust the cook time to 4 minutes.
5. **Pressure Release.** Use the quick-release method to return the pot's pressure to normal.
6. **Finish the dish** Unlock and open the cooker. Stir in the chopped chicken. Cover loosely and set aside for a couple of minutes to warm through.

Carrot, Potato, and Leek Soup

PREP: 10 MINUTES • PRESSURE: HIGH • TIME UNDER PRESSURE: 10 MINUTES • RELEASE: NATURAL

SERVES: 4-6

Ingredients

1 tablespoon olive oil

2 tablespoons unsalted butter

1 medium leek, white and pale green parts only, coarsely chopped

2 teaspoons salt

1 pound carrots, coarsely chopped

1 large potato, peeled and coarsely chopped

4 cups salt-free Chicken Stock

Freshly ground black pepper

1 bouquet garni (parsley sprigs, bay leaf, sprig of thyme, tied tightly with string or in cheesecloth)

¼ cup heavy cream

⅛ teaspoon freshly grated nutmeg

Fresh thyme sprigs or chopped fresh chives, for serving

Directions

1. **Preparing the Ingredients.** Heat the COSORI Pressure Cooker Multi-Cooker ® using the "Sauté" function, add the oil and butter, and cook until the butter has melted. Stir in the chopped leeks and salt and sauté, stirring infrequently, until the leeks have softened, about 5 minutes. Add the carrots, and cook, stirring infrequently, until they are golden on one side, about 5 more minutes. Add the potato, stock, pepper to taste, and the bouquet garni.

2. **High pressure for 10 minutes.** Lock the lid on the COSORI Pressure Cooker Multi-Cooker® and then cook for 10 minutes. To get 10-minutes cook time, press "Manual" Button and use the TIME ADJUSTMENT button to adjust the cook time to 10 minutes.

3. **Pressure Release** When the time is up, open the cooker with the Normal Release method.

4. **Finish the dish.** Fish out and discard the bouquet garni. Using an immersion blender, puree the soup in the cooker. Stir in the cream and nutmeg. Ladle into bowls and dot each serving with a thyme sprig or a few chopped chives.

Colombian Chicken Soup

PREP: 5 MINUTES • PRESSURE: HIGH • TIME UNDER PRESSURE: 17 MINUTES • RELEASE: QUICK

SERVES: 4

Ingredients

- 1 medium yellow onion, cut in half
- 2 medium carrots, cut in half crosswise
- 2 ribs celery, cut in half crosswise
- 3 bone-in chicken breasts (about 2 pounds, or 907 g)
- 5 cups (1.2 L) water
- 1 1/2 teaspoons kosher salt
- 1 1/2 pounds (680 g) Yukon gold potatoes, cut into 1/2-inch (13 mm) pieces
- 1 ear corn, cut into 4 pieces
- 1/4 teaspoon freshly ground black pepper
- 1 avocado
- 1/4 cup (60 g) sour cream
- 1 tablespoon (9 g) capers, rinsed
- 1 teaspoon dried oregano
- 8 sprigs fresh cilantro
- 1 lime, quartered

Directions

1. **Preparing the Ingredients** To the COSORI Pressure Cooker Multi-Cooker ®, add the onion, carrots, celery, chicken, water, and salt.
2. **High pressure for 15 minutes.** Lock the lid on the COSORI Pressure Cooker Multi-Cooker® and then cook for 15 minutes. To get 15-minutes cook time, press "Poultry" Button and then adjust button.
3. **Pressure Release.** Use the "Quick Release" method to vent the steam, then open the lid. Transfer the chicken to a large bowl. When cool enough to handle, shred into pieces, discarding the skin and bones.
   Discard the onion, carrots, and celery. Add the potatoes and corn to the broth.
4. **High pressure for 2 minutes.** Lock the lid on the COSORI Pressure Cooker Multi-Cooker® and then cook for 2 minutes. To get 2-minutes cook time, press "Manual" button and use the TIME ADJUSTMENT button to adjust the cook time to 2 minutes.
5. **Pressure Release.** Use the "Quick Release" method to vent the steam, then open the lid.
6. **Finish the dish.** Stir in the chicken and pepper.
   Divide the soup among bowls. Peel, pit, and slice the avocado. Top the soup with the avocado, sour cream, capers, oregano, and cilantro.
   Serve with the lime quarters for squeezing.
   Enjoy!

Main Dishes – Seafood

Farro With Fennel And Smoked Trout

PREP: 5 MINUTES • PRESSURE: 17 MINUTES • TOTAL: 22 MINUTES • PRESSURE LEVEL: HIGH • RELEASE: QUICK
SERVES 4

Ingredients

1 cup semi-perlato farro
1 large fennel bulb, trimmed and shaved into thin strips
½ cup regular or low-fat mayonnaise
¼ cup regular or low-fat sour cream
3 tablespoons lemon juice
2 tablespoons Dijon mustard
1 teaspoon sugar
1 teaspoon ground black pepper
12 ounces smoked trout, skinned and chopped

Directions

1. **Preparing the Ingredients.** Pour the farro into the COSORI Pressure Cooker Multi-Cooker®; pour in enough water that the grains are submerged by 2 inches.
2. **High pressure for 17 minutes.** Lock the lid on the COSORI Pressure Cooker Multi-Cooker® and then cook for 17 minutes. To get 17-minutes cook time, press "Manual" Button and use the TIME ADJUSTMENT button to adjust the cook time to 17 minutes.
3. **Pressure Release.** Use the quick-release method to drop the pot's pressure to normal.
4. **Finish the dish.** Unlock and open the cooker. Place the fennel strips in a colander set in the sink and drain the farro into the colander over the fennel. Toss well, then let cool for 30 minutes in the colander.
   Whisk the mayonnaise, sour cream, lemon juice, mustard, sugar, and pepper in a large serving bowl until creamy. Add the farro, fennel, and smoked trout; toss gently to coat well.

Pasta with Tuna and Capers

PREP: 2 MINUTES • PRESSURE: 3 MINUTES • TOTAL: 5 MINUTES • PRESSURE LEVEL: HIGH • RELEASE: QUICK
SERVES 2-4

Ingredients

1 tablespoon olive oil
1 garlic clove
3 anchovies
2 cups tomato puree
1½ teaspoons salt
16 oz. (500g) fusilli pasta
2 5.5oz (160g) cans Tuna packed in olive oil water to cover
2 tablespoons capers

Directions

1. **Preparing the Ingredients.** In the pre-heated COSORI Pressure Cooker Multi-Cooker® on "Sauté" mode, add the oil, garlic and anchovies. Sauté until the anchovies begin to disintegrate and the garlic cloves are just starting to turn golden.
   Add the tomato puree and salt and mix together.
   Pour in the un-cooked pasta, and the contents of one tuna can (5 oz.) mixing to coat the dry pasta evenly.
   Flatten the pasta in an even layer and pour in just enough water to cover.
2. **High pressure for 3 minutes.** Lock the lid on the COSORI Pressure Cooker Multi-Cooker® and then cook for 3 minutes. To get 3-minutes cook time, press "Manual" button and use the TIME ADJUSTMENT button to adjust the cook time to 3 minutes.
3. **Pressure Release** When time is up, open the cooker by releasing the pressure.
4. **Finish the dish** Mix in the last 5oz of tuna and sprinkle with capers before serving.
5. Enjoy!

Shrimp And Tomatillo Casserole

PREP: 10 MINUTES • PRESSURE: 9 MINUTES • TOTAL: 20 MINUTES • PRESSURE LEVEL: HIGH • RELEASE: QUICK
SERVES 4

Ingredients

    2 tablespoons olive oil
    1 medium yellow onion, chopped
    1 small fresh jalapeño chile, stemmed, seeded, and minced
    2 teaspoons minced garlic
    1½ pounds fresh tomatillos, husked and chopped
    ½ cup bottled clam juice
    2 tablespoons fresh lime juice
    1½ pounds medium shrimp (about 30 per pound), peeled and deveined
    ¼ cup loosely packed fresh cilantro leaves, chopped
    1 cup shredded Monterey jack cheese (about 4 ounces)

Directions

1. **Preparing the Ingredients.** Heat the oil in the COSORI Pressure Cooker Multi-Cooker® turned to the "Browning" function. Add the onion and cook, stirring often, until translucent, about 3 minutes.

   Add the jalapeño and garlic; cook until aromatic, stirring all the while, less than a minute.

   Stir in the tomatillos, clam juice, and lime juice.

2. **High pressure for 9 minutes.** Lock the lid on the COSORI Pressure Cooker Multi-Cooker® and then cook for 9 minutes. To get 9-minutes cook time, press "Manual" button and use the TIME ADJUSTMENT button to adjust the cook time to 9 minutes.

3. **Pressure Release** Use the quick-release method.

4. **Finish the dish.** Unlock and open the pot. Turn the COSORI Pressure Cooker Multi-Cooker® to its "browning" or "simmer" function. Stir in the shrimp and cilantro; cook for 2 minutes, stirring frequently. Sprinkle the cheese over the top of the casserole, cover the cooker, and lock the lid in place. Set aside off the heat for 2 minutes to melt the cheese and blend the flavors.

   Use the quick-release method (if necessary) to bring any pressure in the pot back to normal.

5. Unlock and open the pot. Stir gently before serving.

Lemon and Dill Fish Packets

PREP: 10 MINUTES • PRESSURE: 5 MINUTES • TOTAL: 15 MINUTES • PRESSURE LEVEL: HIG • RELEASE: QUICK
SERVES 2

Ingredients

2 tilapia or cod fillets
Salt, pepper, and garlic powder
2 sprigs fresh dill
4 slices lemon
2 tablespoons butter

Directions

1. **Preparing the Ingredients.** Lay out 2 large squares of parchment paper.
   Place a fillet in the center of each parchment square, and then season with a generous amount of salt, pepper, and garlic powder.
   On each fillet, place in order: 1 sprig of dill, 2 lemon slices, and 1 tablespoon of butter.
   For best results, place a small metal rack or trivet at the bottom of your COSORI Pressure Cooker Multi-Cooker®.
   Pour 1 cup of water into the cooker to create a water bath.
   Close up parchment paper around the fillets, folding to seal, and then place both packets on metal rack inside cooker.
2. **High pressure for 5 minutes**. Lock the lid on the COSORI Pressure Cooker Multi-Cooker® and then cook for 5 minutes. To get 5-minutes cook time, press "Beans/Chili" button.
3. **Pressure Release** Perform a quick release to release the cooker's pressure. Unwrap packets and serve.
   There is no need to remove the fish from the packets before serving. In fact, it makes a really nice presentation.

Fish Filets

PREP: 5 MINUTES • PRESSURE: 5 MINUTES • TOTAL: 10 MINUTES • PRESSURE LEVEL: LOW • RELEASE: NORMAL
SERVES 2

Ingredients

4 White Fish fillets (any white fish)
1 lb. (500g) Cherry Tomatoes, halved
1 cup Black salt-cured Olives (Taggiesche, French or Kalamata)
2 Tbsp.Pickled Capers
1 bunch of fresh Thyme Olive Oil
1 clove of garlic, pressed
Salt and pepper to taste

Directions

1. **Preparing the Ingredients.** Prepare the base of the COSORI Pressure Cooker Multi-Cooker® with 1½ to 2 cups of water and trivet or steamer basket.
   Line the bottom of the heat-proof bowl with cherry tomato halves (to keep the fish filet from sticking), add Thyme (reserve a few springs for garnish).
   Place the fish fillets over the cherry tomatoes, sprinkle with remaining tomatoes, crushed garlic, a dash of olive oil and a pinch of salt.
   Insert the dish in the COSORI Pressure Cooker Multi-Cooker® - if your heat proof dish does not have handles construct them by making a long aluminum sling.
2. **Low pressure for 5 minutes.** Lock the lid on the COSORI Pressure Cooker Multi-Cooker® and then cook for 5 minutes. To get 5-minutes cook time, press "Beans/Chili" button.
3. **Pressure Release** Perform a quick release to release the cooker's pressure.
   Distribute fish into individual plates, top with cherry tomatoes, and sprinkle with olives, capers, fresh Thyme, a crackle of pepper and a little swirl of fresh olive oil.

Per Serving Calories: 278.2; Fat: 5.8g; Carbohydrates: 18.8g; Sodium: 1056.8mg; Fiber: 2.5g; Protien: 25.6g

Mediterranean Tuna Noodle Delight

PREP: 6 MINUTES • PRESSURE: 10 MINUTES • TOTAL: 16 MINUTES • PRESSURE LEVEL: HIGH • RELEASE: NATURAL
SERVES 2

Ingredients

1 Tablespoon of Oil
½ cup of chopped red onion
8 ounces of dry wide egg noodles (uncooked)
1 can (14 ounces) diced tomatoes with basil, garlic and oregano(undrained) or any kind you have on hand.
1-1/4 cups of water
¼ teaspoon of salt
1/8 teaspoon of pepper
1 can of tuna fish in water, drained
1 jar (7.5 oz.) marinated artichoke hearts, drained with saving the liquid, then chop it up
Crumpled feta cheese
Fresh chopped parsley or dried

Directions

1. **Preparing the Ingredients** Sauté the red onion for about 2 minutes.
   Add the dry noodles, tomatoes, water, salt and pepper .
2. **High pressure for 10 minutes.** Lock the lid on the COSORI Pressure Cooker Multi-Cooker® and then cook for 10 minutes. To get 10-minutes cook time, press "Manual" button.
3. **Pressure Release.** Release the pressure using natural release method.
   Turn off the warm setting.
   Add tuna, artichokes and your reserved liquid from the artichokes and sauté on normal while stirring for about 4 more minutes till hot.
   Plate, then top with a little feta cheese and parsley to your liking.

Per Serving Calories: 258.3; Fat: 5.8g; Carbohydrates: 15.8g; Sugar: 0.2g; Sodium: 1146.8mg; Fiber: 2.5g; Protien: 29.6g

Beer Potato Fish

PREP: 15 MINUTES • PRESSURE: 40 MINUTES • TOTAL: 55 MINUTES • PRESSURE LEVEL: LOW • RELEASE: NATURAL

SERVES 6

Ingredients

- 1 pound fish fillet
- 4 medium size potatoes, peeled and diced
- 1 cup beer
- 1 red pepper sliced
- 1 tablespoon oil
- 1 tablespoon oyster flavored sauce
- 1 tablespoon rock candy
- 1 teaspoon salt

Directions

1. **Preparing the Ingredients** .Put all ingredients into your COSORI Pressure Cooker Multi-Cooker®.
2. **High pressure for 40 minutes.** Lock the lid on the COSORI Pressure Cooker Multi-Cooker® and then cook for 40 minutes. To get 40-minutes cook time, press "Manual" button and use the TIME ADJUSTMENT button to adjust the cook time to 40 minutes.
3. **Pressure Release.** Release the pressure using natural release method
   Then that is it! Simple, fast, delicious, retaining flavour and nutrition, consistent results all the time.
4. Serve and Enjoy!

Per Serving Calories: 250.3; Fat: 4.8g; Sodium: 1146.8mg; Fiber: 2.5g; Protien: 25.6g

Main Dishes – Vegetables

CHICKPEA STEW WITH CARROTS, DATES, AND CRISP ARTICHOKES

PREP: 5 MINUTES • PRESSURE: 12 MINUTES • TOTAL: 17 MINUTES • PRESSURE LEVEL: HIGH • RELEASE: QUICK

SERVES 4

Ingredients

- 1½ cups dried chickpeas
- 2 cups chicken broth
- 2 tablespoons all-purpose flour
- 2½ tablespoons olive oil
- 1 medium red onion, halved and sliced into thin half-moons
- 2 teaspoons minced garlic
- 1 tablespoon sweet paprika
- ½ teaspoon ground cinnamon
- ½ teaspoon ground coriander
- ½ teaspoon ground cumin
- ½ teaspoon salt
- One 14-ounce can diced tomatoes (about 1¾ cups)
- 1 pound "baby" carrots, cut into 1-inch pieces
- 6 pitted dates, preferably Medjool, chopped
- One 9-ounce box frozen artichoke heart quarters, thawed and squeezed of excess moisture

Directions

1. **Preparing the Ingredients.** Soak the chickpeas in a big bowl of water for at least 12 hours or up to 16 hours.

   Drain the chickpeas in a colander set in the sink. Whisk the broth and flour in a medium bowl until the flour dissolves.

   Heat 1½ tablespoons oil in the COSORI Pressure Cooker Multi-Cooker® turned to the browning function. Add the onion and cook, stirring often, until softened, about 4 minutes.

   Stir in the garlic, paprika, cinnamon, coriander, cumin, and salt until aromatic, about 30 seconds. Pour in the tomatoes as well as the broth mixture. Stir well, then add the carrots, dates, and drained chickpeas.

2. **High pressure for 12 minutes.** Lock the lid on the COSORI Pressure Cooker Multi-Cooker® and then cook for 12 minutes. To get 12-minutes cook time, press "Manual" Button and use the TIME ADJUSTMENT button to adjust the cook time to 12 minutes.

3. **Pressure Release** Use the quick-release method to drop the pot's pressure back to normal.

4. **Finish the dish.** Unlock and open the cooker. Heat the remaining tablespoon oil in a large nonstick skillet set over medium-high heat. Add the artichoke heart quarters;

fry until brown and crisp, stirring and turning occasionally, about 10 minutes. Dish up the chickpea mixture into big bowls and top with the crisp artichoke bits.

Ratatouille

PREP: 5 MINUTES • PRESSURE: 4 MINUTES • TOTAL: 9 MINUTES • PRESSURE LEVEL: HIGH • RELEASE: QUICK
SERVES 4

Ingredients

Kosher salt, for salting and seasoning
1 small eggplant, peeled and sliced ½ inch thick
1 medium zucchini, sliced ½ inch thick
2 tablespoons olive oil
1 cup chopped onion
3 garlic cloves, minced or pressed
1 small green bell pepper, cut into ½-inch chunks (about 1 cup)
1 small red bell pepper, cut into ½-inch chunks (about 1 cup)
1 rib celery, sliced (about 1 cup)
1 (14.5-ounce) can diced tomatoes, undrained
¼ cup water
½ teaspoon dried oregano
¼ teaspoon freshly ground black pepper
2 tablespoons minced fresh basil
¼ cup pitted green or black olives (optional)

Directions

1. **Preparing the Ingredients.** Place a rack over a baking sheet. With kosher salt, very liberally salt one side of the eggplant and zucchini slices, and place them, salted-side down, on the rack. Salt the other side. Let the slices sit for 15 to 20 minutes, or until they start to exude water (you'll see it beading up on the surface of the slices and dripping into the sheet pan). Rinse the slices, and blot them dry. Cut the zucchini slices into quarters and the eggplant slices into eighths.

   Turn the COSORI Pressure Cooker Multi-Cooker® to "brown," heat the olive oil until it shimmers and flows like water. Add the onion and garlic, and sprinkle with a pinch or two of kosher salt. Cook for about 3 minutes, stirring, until the onions just begin to brown.

   Add the eggplant, zucchini, green bell pepper, red bell pepper, celery, and tomatoes with their juice, water, and oregano.

2. **High pressure for 4 minutes.** Lock the lid on the COSORI Pressure Cooker Multi-Cooker® and then cook for 4 minutes. To get 4-minutes cook time, press "Manual" button and use the TIME ADJUSTMENT button to adjust the cook time to 4 minutes.

3. **Pressure Release.** Use the quick-release method.

4. **Finish the dish.** Unlock and remove the lid. Stir in the pepper, basil, and olives (if using). Taste, adjust the seasoning as needed, and serve.

   While this vegetable dish is usually served on its own, it's great tossed with cooked pasta or served over polenta.

PER SERVING: CALORIES: 149; FAT: 8G; SODIUM: 55MG; CARBOHYDRATES: 20G; FIBER: 8G; PROTEIN: 4G

Beets *and* Greens *with* Horseradish Sauce

PREP: 5 MINUTES • PRESSURE: 10 MINUTES • TOTAL: 15 MINUTES • PRESSURE LEVEL: HIGH • RELEASE: NATURAL
SERVES 4

Ingredients

2 large or 3 small beets with greens, scrubbed and root ends trimmed
1 cup water, for steaming
2 tablespoons sour cream
1 tablespoon whole milk
1 teaspoon prepared horseradish
¼ teaspoon lemon zest
⅛ teaspoon kosher salt, divided
2 teaspoons unsalted butter
1 tablespoon minced fresh chives

Directions

1. **Preparing the Ingredients.** Trim off the beet greens and set aside. If the beets are very large (3 inches or more in diameter), quarter them; otherwise, halve them. Add the water and insert the steamer basket or trivet. Place the beets on the steamer insert.

2. **High pressure for 10 minutes.** Lock the lid on the COSORI Pressure Cooker Multi-Cooker® and then cook for 10 minutes. To get 10-minutes cook time, press "Manual" Button and use the TIME ADJUSTMENT button to adjust the cook time to 10 minutes
   When the timer goes off, turn the cooker off. ("Warm" setting, turn off).

3. **Pressure Release** Let the pressure to come down naturally .

   While the beets are cooking and the pressure is releasing, wash the greens and slice them into ½-inch-thick ribbons, removing any tough stems. In a small bowl, whisk together the sour cream, milk, horseradish, lemon zest, and $1/16$ teaspoon of kosher salt.

4. **Finish the dish.** When the pressure has released completely, unlock and remove the lid. Remove the beets and cool slightly; then use a paring knife or peeler to peel them. Slice them into large bite-size pieces and set aside.
   Remove the steamer from the COSORI Pressure Cooker Multi-Cooker®, and pour out the water. Turn the COSORI Pressure Cooker Multi-Cooker® to "brown." Add the butter to melt. When the butter stops foaming, add the beet greens and sprinkle with the remaining $1/16$ teaspoon of kosher salt. Cook for 3 to 4 minutes, stirring, until wilted. Return the beets to the COSORI Pressure Cooker Multi-Cooker® and heat for 1 or 2 minutes, stirring. Transfer the beets and greens to a platter, and drizzle with the sour cream mixture. Sprinkle with the chives, and serve.
    It may be tempting to cool the beets completely before you peel them, but that would be a mistake. Beets are easiest to peel when they're just cool enough to handle; if they get too cool, the skins tend to stick.

PER SERVING: CALORIES: 70; FAT: 4G; SODIUM: 162MG; CARBOHYDRATES: 9G; FIBER: 2G; PROTEIN: 2G

WARM QUINOA AND POTATO SALAD

PREP: 5 MINUTES • PRESSURE: 10 MINUTES • TOTAL: 15 MINUTES • PRESSURE LEVEL: HIGH • RELEASE: QUICK

SERVES 6

Ingredients

¼ cup white balsamic vinegar

1 tablespoon Dijon mustard

1 teaspoon sweet paprika

½ teaspoon ground black pepper

¼ teaspoon celery seeds

¼ teaspoon salt

¼ cup olive oil

1½ pounds tiny white potatoes, halved

1 cup blond (white) quinoa

1 medium shallot, minced

2 medium celery stalks, thinly sliced

1 large dill pickle, diced

Directions

1. **Preparing the Ingredients.** Whisk the vinegar, mustard, paprika, pepper, celery seeds, and salt in a large serving bowl until smooth; whisk in the olive oil in a thin, steady stream until the dressing is fairly creamy.

   Place the potatoes and quinoa in the COSORI Pressure Cooker Multi-Cooker®; add enough cool tap water so that the ingredients are submerged by 3 inches (some of the quinoa may float).

2. **High pressure for 10 minutes.** Lock the lid on the COSORI Pressure Cooker Multi-Cooker® and then cook for 10 minutes. To get 10-minutes cook time, press "Manual" Button and use the TIME ADJUSTMENT button to adjust the cook time to 10 minutes.

3. **Pressure Release** Use the quick-release method to bring the pot's pressure back to normal.

4. **Finish the dish.** Unlock and open the pot. Drain the contents of the pot into a colander lined with paper towels or into a fine-mesh sieve in the sink. Do not rinse. Transfer the potatoes and quinoa to the large bowl with the dressing. Add the shallot, celery, and pickle; toss gently and set aside for a minute or two to warm up the vegetables.

Buttery Carrots With Pancetta

PREP: 5 MINUTES • PRESSURE: 7 MINUTES • TOTAL: 12 MINUTES • PRESSURE LEVEL: HIGH • RELEASE: QUICK
SERVES 4 - 6

Ingredients

4 ounces pancetta, diced
1 medium leek, white and pale green parts only, sliced lengthwise, washed, and thinly
sliced
¼ cup moderately sweet white wine, such as a dry Riesling
1 pound baby carrots
½ teaspoon ground black pepper
2 tablespoons unsalted butter, cut into small bits

Directions

1. **Preparing the Ingredients.** Put the pancetta in the COSORI Pressure Cooker Multi-Cooker® turned to the "browning" function. Fry until crisp and well browned, stirring occasionally, about 3 minutes.
   Add the leek; cook, stirring often, until softened, about 1 minute. Pour in the wine and scrape up any browned bits at the bottom of the pot as it comes to a simmer.
   Add the carrots and pepper; stir well. Scrape and pour the contents of the COSORI Pressure Cooker Multi-Cooker® into a 1-quart, round, high-sided soufflé or baking dish. Dot with the bits of butter. Lay a piece of parchment paper on top of the dish, then a piece of aluminum foil. Seal the foil tightly over the baking dish.
   Set the COSORI Pressure Cooker Multi-Cooker® rack inside, and pour in 2 cups water. Use aluminum foil to build a sling for the baking dish; lower the baking dish into the cooker.
2. **High pressure for 7 minutes.** Lock the lid on the COSORI Pressure Cooker Multi-Cooker® and then cook for 7 minutes. To get 7-minutes cook time, press "Manual" button and use the TIME ADJUSTMENT button to adjust the cook time to 7 minutes.
3. **Pressure Release.** Use the quick-release method to return the pot's pressure to normal.
4. **Finish the dish.** Unlock and open the pot. Use the foil sling to lift the baking dish out of the cooker. Uncover, stir well, and serve.

Braised Red Cabbage With Apples

PREP: 5 MINUTES • PRESSURE: 13 MINUTES • TOTAL: 18 MINUTES • PRESSURE LEVEL: HIGH • RELEASE: QUICK
SERVES 4

Ingredients
4 thin bacon slices, chopped
1 small red onion, chopped
1 medium tart green apple, such as Granny Smith, peeled, cored, and chopped
1 teaspoon dried thyme
¼ teaspoon ground allspice
¼ teaspoon ground mace
1 tablespoon packed dark brown sugar
1 tablespoon balsamic vinegar
1 medium red cabbage (about 2 pounds), cored and thinly sliced
½ cup chicken broth

Directions
1. **Preparing the Ingredients.** Fry the bacon in the COSORI Pressure Cooker Multi-Cooker® turned to the "Browning" function, stirring often, until crisp, about 4 minutes. Add the onion to the pot; cook, stirring often, until soft, about 4 minutes. Add the apple, thyme, allspice, and mace. Cook about 1 minute, stirring all the while, until fragrant. Stir in the brown sugar and vinegar; keep stirring until bubbling, about 1 minute.
Add the cabbage; toss well to mix evenly with the other ingredients. Drizzle the broth over the cabbage mixture.
2. **High pressure for 13 minutes.** Lock the lid on the COSORI Pressure Cooker Multi-Cooker® and then cook for 13 minutes. To get 13-minutes cook time, press "Manual" Button, and use the TIME ADJUSTMENT button to adjust the cook time to 13 minutes.
3. **Pressure Release.** Use the quick-release method to return the pot to normal pressure. Unlock and open the pot. Stir well before serving.

SAGE-BUTTER SPAGHETTI SQUASH

PREP: 5 MINUTES • PRESSURE: 12 MINUTES • TOTAL: 17 MINUTES • PRESSURE LEVEL: HIGH • RELEASE: QUICK
SERVES 6

Ingredients

One 3- to 3½-pound spaghetti squash, halved lengthwise and seeded
6 tablespoons unsalted butter
2 tablespoons packed fresh sage leaves, minced
½ teaspoon salt
½ teaspoon ground black pepper
½ cup finely grated Parmesan cheese (about 1 ounce)

Directions

1. **Preparing the Ingredients.** Put the squash cut side up in the cooker; add 1 cup water.
2. **High pressure for 12 minutes.** Lock the lid on the COSORI Pressure Cooker Multi-Cooker® and then cook for 12 minutes. To get 12-minutes cook time, press "Manual" Button, and use the TIME ADJUSTMENT button to adjust the cook time to 12 minutes.
3. **Pressure Release** Use the quick-release method to bring the pot's pressure back to normal.
4. **Finish the dish.** Unlock and open the cooker. Transfer the squash halves to a cutting board; cool for 10 minutes. Discard the liquid in the cooker. Use a fork to scrape the spaghetti-like flesh off the skin and onto the cutting board; discard the skins.
   Melt the butter in the electric cooker turned to its browning function. Stir in the sage, salt, and pepper, then add all of the squash. Stir and toss over the heat until well combined and heated through, about 2 minutes. Add the cheese, toss well, and serve.

BUTTERY RYE BERRY AND CELERY ROOT SALAD

PREP: 5 MINUTES • PRESSURE: 40 MINUTES • TOTAL: 45 MINUTES • PRESSURE LEVEL: HIGH • RELEASE: QUICK
SERVES 6

Ingredients

¾ cup rye berries
1 medium celeriac (celery root), peeled and shredded through the large holes of a box grater
2 tablespoons unsalted butter
2 tablespoons honey
2 tablespoons apple cider vinegar
½ teaspoon salt
½ teaspoon ground black pepper

Directions

1. **Preparing the Ingredients.** Place the rye berries in the COSORI Pressure Cooker Multi-Cooker®; pour in enough cool tap water so the grains are submerged by 2 inches.
2. **High pressure for 40 minutes.** Lock the lid on the COSORI Pressure Cooker Multi-Cooker® and then cook for 40 minutes. To get 40-minutes cook time, press "Manual" button and use the TIME ADJUSTMENT button to adjust the cook time to40 minutes.
3. **Pressure Release** Use the quick-release method to bring the pot's pressure back to normal.
4. **Finish the dish.** Unlock and open the cooker. Stir in the shredded celeriac. Cover the pot without locking it and set aside for 1 minute. Drain the pot into a large colander set in the sink. Wipe out the cooker.
   Melt the butter in the COSORI Pressure Cooker Multi-Cooker®; turned to its browning function. Add the honey and cook for 1 minute, stirring constantly. Add the drained rye berries and celeriac; cook, stirring constantly, for 1 minute. Stir in the vinegar, salt, and pepper to serve.

Vegetable Stew with Barley

PREP: 10 MINUTES • PRESSURE: 35 MINUTES • TOTAL: 55 MINUTES • PRESSURE LEVEL: HIGH • RELEASE: QUICK
SERVES 6

Ingredients

    6 tomatoes, diced
    2 large carrots, cut into bite size pieces
    3 potatoes cut into chunks
    4 celery stalks cut into bite size pieces
    2 cups of sliced white mushrooms
    1 large onion, diced
    6 cups vegetable stock (or beef/chicken stock)
    ½ cup red wine or rice wine (red wine is preferred)
    1 cup pearl barley
    3 gloves garlic, minced
    1 tablespoon dried parsley flakes
    1 tablespoon dried thyme
    1 bay leaf

Directions

1. **Preparing the Ingredients.** In a nonstick pan add a drizzle of olive oil and quickly sauté the white mushrooms with the minced garlic and onions until golden brown (2-3 minutes on medium heat) then add in the red wine and cook for another minute. Set aside.
   In the COSORI Pressure Cooker Multi-Cooker® add the rest of the ingredients not including the barley.
2. **High pressure for 20 minutes.** Lock the lid on the COSORI Pressure Cooker Multi-Cooker® and then cook for 20 minutes. To get 20-minutes cook time, press "Manual" button.
3. **Pressure Release** Use the quick-release method to bring the pot's pressure back to normal.
   Add in the mushrooms and barley, give it a good stir and add 2 pinches of salt and pepper.
4. **High pressure for 10 minutes.** Lock the lid on the COSORI Pressure Cooker Multi-Cooker® and then cook for 10 minutes. To get 10-minutes cook time, press "Manual" Button and use the TIME ADJUSTMENT button to adjust the cook time to10 minutes.

5. **Pressure Release** Use the quick-release method to bring the pot's pressure back to normal.
6. **Finish the dish** At this point the potatoes and carrots should have soften. Add salt and pepper to taste.
   Serve with your favorite pasta dish fresh baked biscuits.

One-Pot Pasta Puttanesca

PREP: 5 MINUTES • PRESSURE: 8 MINUTES • TOTAL: 13 MINUTES • PRESSURE LEVEL: HIGH • RELEASE: QUICK
SERVES 4

Ingredients

    2 tablespoons olive oil
    1 small red onion, chopped
    1 tablespoon drained and rinsed capers, minced
    1 tablespoon minced garlic
    1 pound eggplant (about 1 large), stemmed and diced (no need to peel)
    2 medium yellow bell peppers, stemmed, cored, and chopped
    One 28-ounce can diced tomatoes (about 3½ cups)
    1¼ cups vegetable broth
    2 tablespoons canned tomato paste
    2 teaspoons dried rosemary
    1 teaspoon dried thyme
    ½ teaspoon ground black pepper
    8 ounces dried whole wheat ziti

Directions

1. **Preparing the Ingredients.** Heat the oil in the COSORI Pressure Cooker Multi-Cooker® turned to the "Browning" function. Add the onion, capers, and garlic; cook, stirring often, just until the onion first begins to soften, about 2 minutes.
Add the eggplant and bell peppers; cook, stirring often, for 1 minute. Mix in the tomatoes, broth, tomato paste, rosemary, thyme, and pepper, stirring until the tomato paste coats everything. Stir in the ziti until coated.
2. **High pressure for 8 minutes.** Lock the lid on the COSORI Pressure Cooker Multi-Cooker® and then cook for 8 minutes. To get 8-minutes cook time, press "Manual" button and use the TIME ADJUSTMENT button to adjust the cook time to 8 minutes.
3. **Pressure Release.** Use the quick-release method to drop the pressure in the pot back to normal.
4. Unlock and open the cooker. Stir well before serving.

Main Dishes –Rice

Perfect Brown Rice
Long-Grain White Rice
Risotto *with* Peas *and* Shrimp
Risotto With Butternut Squash And Porcini
Barley Risotto with Fresh Spinach
Quinoa Risotto With Bacon
Armenian Rice Pilaf
Wild *and* Brown Rice Pilaf
Wild Rice Salad with Apples
Seafood Risotto
Brown Rice Pilaf With Cashews
Wild Rice With Sweet Potatoes
Brown Rice with Lentils
Tiger Prawn Risotto

Perfect Brown Rice

PREP: 5 MINUTES • PRESSURE: 22 MINUTES • TOTAL: 27 MINUTES • PRESSURE LEVEL: HIGH • RELEASE: NATURAL

SERVES 4

Ingredients

- 1½ cups brown rice
- 2½ cups water
- ½ teaspoon salt
- 1 teaspoon olive oil

Directions

1. **Preparing the Ingredients** Place the rice, water, salt, and oil in the COSORI Pressure Cooker Multi-Cooker® base.
2. **High pressure for 20 minutes.** Lock the lid on the COSORI Pressure Cooker Multi-Cooker® and then cook for 20 minutes. To get 20-minutes cook time, press "Brown Rice" button and use the TIME ADJUSTMENT button to adjust the cook time to 20 minutes.
3. **Pressure Release.** When the time is up, open the COSORI Pressure Cooker Multi-Cooker® with the 10-Minute Natural Release method.
4. Fluff the rice with a fork and serve.

Long-Grain White Rice

PREP: 5 MINUTES • PRESSURE: 22 MINUTES • TOTAL: 27 MINUTES • PRESSURE LEVEL: HIGH • RELEASE: NATURAL
SERVES 4

Ingredients

1½ cups long-grain white rice
3 cups water
½ teaspoon salt
1 teaspoon vegetable oil or unsalted butter

Directions

1. **Preparing the Ingredients.** Place the rice, water, salt, and oil in the COSORI Pressure Cooker Multi-Cooker® base.
2. **High pressure for 3 minutes.** Lock the lid on the COSORI Pressure Cooker Multi-Cooker® and then cook for 4 minutes. To get 3-minutes cook time, press "White Rice" button.
3. **Pressure Release.** When the time is up, open the COSORI Pressure Cooker Multi-Cooker® with the 10-Minute Natural Release method.
4. Fluff the rice with a fork and serve.

Risotto *with* Peas *and* Shrimp

PREP: 5 MINUTES • PRESSURE: 6 MINUTES • TOTAL: 11 MINUTES • PRESSURE LEVEL: HIGH • RELEASE: QUICK
SERVES 4

Ingredients

1 tablespoon unsalted butter

½ cup chopped onion

1 cup Arborio rice

⅓ cup white wine

2¾ cups Chicken Stock or low-sodium broth, divided

½ pound raw medium shrimp, shelled and deveined

½ cup frozen peas, thawed

¼ cup grated Parmigiano-Reggiano or similar cheese

Directions

1. **Preparing the Ingredients.** Turn the COSORI Pressure Cooker Multi-Cooker® to "brown," heat the butter until it stops foaming. Add the onion, and cook for about 2 minutes, stirring, until soft. Add the rice, and stir to coat with the butter. Cook for 1 minute, stirring. Stir in the white wine, and cook for 1 to 2 minutes, or until it's almost evaporated. Add 2½ cups of Chicken Stock, and stir to make sure no rice is sticking to the bottom of the cooker.
2. **High pressure for 6 minutes.** Lock the lid on the COSORI Pressure Cooker Multi-Cooker® and then cook for 6 minutes. To get 6-minutes cook time, press "White Rice" button and use the TIME ADJUSTMENT button to adjust the cook time to 6 minutes.
3. **Pressure Release** Use the quick-release method.
4. **Finish the dish.** Unlock and remove the lid. Turn the COSORI Pressure Cooker Multi-Cooker® to "brown," Continue to cook the rice, stirring, for 1 to 2 minutes more, or until the rice is firm just in the very center of the grain and the liquid has thickened slightly. Add the shrimp and peas, and continue to cook for about 4 minutes more, or until the shrimp are cooked. Stir in the Parmigiano-Reggiano. If the risotto is too thick, stir in a little of the remaining ¼ cup of Chicken Stock to loosen it up. Serve immediately.

   Risotto is one of those dishes that lend themselves to almost endless variation. Leftover ham is a great addition, as is smoked salmon or trout, or go vegetarian with "Sautéed" Mushrooms, roasted peppers, or even beets.

PER SERVING: CALORIES: 343; FAT: 6G; SODIUM: 292MG; CARBOHYDRATES: 45G; FIBER: 3G; PROTEIN: 21G

Risotto With Butternut Squash And Porcini

PREP: 5 MINUTES • PRESSURE: 10 MINUTES • TOTAL: 15 MINUTES • PRESSURE LEVEL: HIGH • RELEASE: QUICK

SERVES 6

Ingredients

2 tablespoons unsalted butter

1 medium leek, white and pale green parts only, halved lengthwise, washed, and thinly sliced

1½ cups white Arborio rice

¼ cup dry vermouth

4 cups (1 quart) vegetable broth

2 cups seeded, peeled, and finely chopped butternut squash

½ ounce dried porcini mushrooms, crumbled

1 teaspoon dried thyme

¼ teaspoon saffron threads

½ cup finely grated Parmesan cheese (about 1 ounce)

Directions

1. **Preparing the Ingredients**. Melt the butter in the COSORI Pressure Cooker Multi-Cooker® turned to the "browning" function. Add the leek and cook, stirring often, until softened, about 2 minutes.

   Add the rice; stir until coated in the butter. Pour in the vermouth; stir over the heat until fully absorbed into the grains, 1 to 2 minutes. Add the broth, squash, dried porcini, thyme, and saffron.

2. **High pressure for 10 minutes.** Lock the lid on the COSORI Pressure Cooker Multi-Cooker® and then cook for 10 minutes. To get 10-minutes cook time, press "Manual" button and use the TIME ADJUSTMENT button to adjust the cook time to10 minutes.

3. **Pressure Release.** Use the quick-release method.

4. **Finish the dish.** Unlock and open the cooker. Turn the COSORI Pressure Cooker Multi-Cooker® to its "browning" function. Bring to a simmer, stirring until thickened, about 2 minutes.

5. Stir in the cheese. Put the lid onto the cooker without locking it in place. Set aside for 5 minutes to melt the cheese and blend the flavors. Stir again before serving.

6. Serve and Enjoy!

Barley Risotto with Fresh Spinach

PREP: 6 MINUTES • PRESSURE: 22 MINUTES • TOTAL: 26 MINUTES • PRESSURE LEVEL: HIGH • RELEASE: QUICK

SERVES 6

Ingredients

1 tablespoon olive oil
1 tablespoon light margarine
1 yellow onion, diced
1 cup pearled barley
4 cups chicken stock or broth
Juice of 1 lemon
1 tablespoon minced garlic
4 cups baby spinach
1/4 cup grated Parmesan cheese
Salt and pepper

Directions

1. **Preparing the Ingredients.** With the cooker's lid off, heat oil and margarine on "brown," until oil is sizzling and margarine is melted.
   Place diced onion in the cooker, and sauté until translucent, 5 minutes.
   Stir in barley, and sauté 1 additional minute.
   Add the chicken broth, lemon juice, and minced garlic.

2. **High pressure for 25 minutes.** Lock the lid on the COSORI Pressure Cooker Multi-Cooker® and then cook for 25 minutes. To get 25-minutes cook time, press "Manual" button and use the TIME ADJUSTMENT button to adjust the cook time to 25 minutes.

3. **Pressure Release.** Let the pressure release naturally 5 minutes before performing a quick release for any remaining pressure.
   With the cooker's lid off, set to "brown," to sauté, and stir in spinach and Parmesan cheese, simmering until spinach cooks down. Season with salt and pepper to taste before serving.

Quinoa Risotto With Bacon

PREP: 7 MINUTES • PRESSURE: 9 MINUTES • TOTAL: 16 MINUTES • PRESSURE LEVEL: HIGH • RELEASE: QUICK

SERVES 4

Ingredients

3 ounces slab bacon, diced

6 medium scallions, thinly sliced

12 cherry tomatoes, halved

¼ cup dry vermouth

3½ cups chicken broth

1½ cups white or red quinoa, rinsed if necessary

3 fresh thyme sprigs

¼ cup finely grated Parmesan cheese (about ½ ounce)

½ teaspoon ground black pepper

Directions

1. **Preparing the Ingredients.** Place the bacon in the COSORI Pressure Cooker Multi-Cooker® turned to the "browning" function. Fry until crisp, stirring occasionally, about 4 minutes.

   Add the scallions; stir over the heat until softened, about 1 minute. Put in the tomatoes; cook just until they begin to break down, about 2 minutes, stirring occasionally. Pour in the vermouth; as it comes to a simmer, scrape up any browned bits in the bottom of the cooker.

   Stir in the broth, quinoa, and thyme sprigs.

2. **High pressure for 9 minutes.** Lock the lid on the COSORI Pressure Cooker Multi-Cooker® and then cook for 9 minutes. To get 9-minutes cook time, press "Manual" button and use the TIME ADJUSTMENT button to adjust the cook time to 9 minutes.

3. **Pressure Release.** Return the pot's pressure to normal with the quick-release method.

4. **Finish the dish.** Unlock and open the cooker. Turn the COSORI Pressure Cooker Multi-Cooker® to its "browning" function. Discard the thyme sprigs. Bring the mixture in the pot to a simmer; cook, stirring often, until thickened, 2 to 3 minutes. Stir in the cheese and pepper to serve.

Armenian Rice Pilaf

PREP: 7 MINUTES • PRESSURE: 3 MINUTES • TOTAL: 10 MINUTES • PRESSURE LEVEL: HIGH • RELEASE: NATURAL
SERVES 4

Ingredients

2 tablespoons unsalted butter
1 tablespoon olive oil
½ cup vermicelli or angel hair pasta, broken into 1-inch pieces
2 cups long-grain white or basmati rice
4 cups salt-free Chicken Stock, preferably double-strength
2 teaspoons salt

Directions

1. **Preparing the Ingredients.** Heat the COSORI Pressure Cooker Multi-Cooker® using the "Sauté" Function, add the butter and oil, and cook until the butter has melted. Add the vermicelli and stir well to coat. Sauté until the pieces just begin to turn golden. Add the rice; stir well to coat and toast for about 1 minute. Add the chicken stock and salt.
2. **High pressure for 3 minutes.** Lock the lid on the COSORI Pressure Cooker Multi-Cooker® and then cook for 3 minutes. To get 3-minutes cook time, press "White Rice" button and use the TIME ADJUSTMENT button to adjust the cook time to 3 minutes.
3. **Pressure Release.** When the time is up, open the COSORI Pressure Cooker Multi-Cooker® with the 10-Minute Natural Release method.
4. Mix the pilaf well, pulling up the rice from the bottom of the COSORI Pressure Cooker Multi-Cooker® to the top before serving.

Wild *and* Brown Rice Pilaf

PREP: 5 MINUTES • PRESSURE: 27 MINUTES • TOTAL: 32 MINUTES • PRESSURE LEVEL: HIGH • RELEASE: COMBINATION

SERVES 4

Ingredients

1 tablespoon olive oil
¾ cup diced onion
1 garlic clove, minced
⅓ cup wild rice
⅔ cup water
½ teaspoon kosher salt, divided, plus additional for seasoning
½ cup brown rice
¾ cup low-sodium vegetable broth
¼ cup dry white wine
1 bay leaf
1 fresh thyme sprig, or ¼ teaspoon dried thyme
2 tablespoons chopped fresh parsley

Directions

1. **Preparing the Ingredients.** Set the COSORI Pressure Cooker Multi-Cooker® to "brown," heat the olive oil until it shimmers and flows like water. Add the onion and garlic, and cook for about 3 minutes, stirring, until the garlic is fragrant and the onions soften and separate. Add the wild rice, water, and ¼ teaspoon of kosher salt, and stir.
2. **High pressure for 15 minutes.** Lock the lid on the COSORI Pressure Cooker Multi-Cooker® and then cook for 15 minutes. To get 15-minutes cook time, press "Poultry" Button and then adjust button.
3. **Pressure Release.** Use the quick-release method.
   Unlock and remove the lid. Stir in the brown rice, vegetable broth, remaining ¼ teaspoon of kosher salt, white wine, bay leaf, and thyme.
4. **High pressure for 12 minutes.** Lock the lid on the COSORI Pressure Cooker Multi-Cooker® and then cook for 12 minutes. To get 12-minutes cook time, press "Brown Rice" button and the adjust button.
   When the timer goes off, turn the cooker off. ("Warm" setting, turn off).
5. **Pressure Release.** After cooking, use the natural method to release pressure for 12 minutes, then the quick method to release the remaining pressure.
6. **Finish the dish.** Unlock and remove the lid. Remove the bay leaf and thyme sprig, and stir in the parsley. Taste and adjust the seasoning, as needed. Replace but *do not lock* the lid. Let the rice steam for about 4 minutes, fluff gently with a fork, and serve.

PER SERVING: CALORIES: 195; FAT: 4G; SODIUM: 309MG; CARBOHYDRATES: 32G; FIBER: 2G; PROTEIN: 5G

Wild Rice Salad with Apples

PREP: 5 MINUTES • PRESSURE: 18 MINUTES • TOTAL: 23 MINUTES • PRESSURE LEVEL: HIGH • RELEASE: NATURAL
SERVES 4

Ingredients
4 cups water
1¼ teaspoons kosher salt, divided
1 cup wild rice
⅓ cup walnut or olive oil
3 tablespoons cider vinegar
¼ teaspoon celery seed
⅛ teaspoon freshly ground black pepper
Pinch granulated sugar
½ cup walnut pieces, toasted
2 or 3 celery stalks, thinly sliced (about 1 cup)
1 medium Gala, Fuji, or Braeburn apple, cored and cut into ½-inch pieces

Directions
1. **Preparing the Ingredients.** Add the water into the COSORI Pressure Cooker Multi-Cooker®, and 1 teaspoon of kosher salt. Stir in the wild rice.
2. **High pressure for 18 minutes.** Lock the lid on the COSORI Pressure Cooker Multi-Cooker® and then cook for 18 minutes. To get 18-minutes cook time, press Meat/Chicken button and use the TIME ADJUSTMENT button to adjust the cook time to 18 minutes.
3. **Pressure Release.** Use the natural-release method.
4. **Finish the dish.** Unlock and remove the lid. The rice grains should be mostly split open. If not, simmer the rice for several minutes more, in the COSORI Pressure Cooker Multi-Cooker® set to "brown," until at least half the grains have split. Drain and cool slightly.

   To a small jar with a tight-fitting lid, add the walnut oil, cider vinegar, celery seed, the remaining ¼ teaspoon of kosher salt, the pepper, and the sugar, and shake until well combined.

   To a medium bowl, add the cooled rice, walnuts, celery, and apple. Pour half of the dressing over the salad, and toss gently to coat, adding more dressing as desired.
5. Serve.

PER SERVING: CALORIES: 335; FAT: 17G; SODIUM: 162MG; CARBOHYDRATES: 37G; FIBER: 5G; PROTEIN: 13G

Seafood Risotto

PREP: 10 MINUTES • PRESSURE: 6 MINUTES • TOTAL: 16 MINUTES • PRESSURE LEVEL: HIGH • RELEASE: NATURAL

SERVES 4

Ingredients

3 cups mixed seafood (shrimp, calamari, clams, etc.)
Water, as needed
2 tablespoons olive oil, plus more to finish
3 garlic cloves, chopped
3 oil-packed anchovies
2 cups Arborio or Carnaroli rice
Freshly squeezed juice of 1 lemon
2 teaspoons salt
¼ teaspoon ground white pepper
1 bunch flat-leaf parsley, chopped
Lemon wedges, for serving

Directions

1. **Preparing the Ingredients.** Separate the shellfish from the other seafood and set the shellfish aside. Add the remaining seafood to a 4-cup measuring cup and add water to just over the 4-cup mark.

   Heat the COSORI Pressure Cooker Multi-Cooker® using the "Sauté" mode, add the oil, and heat briefly. Stir in the garlic and anchovies and sauté until the garlic is golden and the anchovies are broken up. Add the rice, stirring to coat well. While you continue to stir, look carefully at the rice, it will first become wet and look slightly transparent and pearly; then it will slowly begin to look dry and solid white again. At that point pour in the lemon juice. Scrape the bottom of the COSORI Pressure Cooker Multi-Cooker® gently, and keep stirring until all of the juice has evaporated. Stir in the seafood and water and the salt and pepper. Place the shellfish on top without stirring any further.

2. **High pressure for 6 minutes**. Lock the lid on the COSORI Pressure Cooker Multi-Cooker® and then cook for 6 minutes. To get 6-minutes cook time, press "White Rice button and use the TIME ADJUSTMENT button to adjust the cook time to 6 minutes.

3. **Pressure Release** When the time is up, open the COSORI Pressure Cooker Multi-Cooker® with the Natural Release method.

   Stir the risotto. Swirl some oil over the top and sprinkle with parsley. Serve with lemon wedges.

Brown Rice Pilaf With Cashews

PREP: 5 MINUTES • PRESSURE: 33 MINUTES • TOTAL: 38 MINUTES • PRESSURE LEVEL: HIGH • RELEASE: QUICK
SERVES 6

Ingredients

3 tablespoons unsalted butter
1 large leek, white and pale green parts only, halved lengthwise, washed, and thinly sliced
½ teaspoon dried thyme
½ teaspoon salt
⅛ teaspoon ground turmeric
1½ cups long-grain brown rice, such as brown basmati
3 cups vegetable or chicken broth
½ cup chopped roasted unsalted cashews

Directions

1. **Preparing the Ingredients.** Melt the butter in the COSORI Pressure Cooker Multi-Cooker® turned to the "Browning" function. Add the leek and cook, stirring often, until softened, about 2 minutes. Stir in the thyme, salt, and turmeric until fragrant, less than half a minute. Add the rice and cook for 1 minute, stirring all the while. Pour in the broth and stir well to get any browned bits off the bottom of the cooker.
2. **High pressure for 33 minutes.** Lock the lid on the COSORI Pressure Cooker Multi-Cooker® and then cook for 33 minutes. To get 33-minutes cook time, press "Manual" Button and use the TIME ADJUSTMENT button to adjust the cook time to 33 minutes.
3. **Pressure Release.** Use the quick-release method to return the pot's pressure to normal but do not open the cooker. Set aside for 10 minutes to steam the rice.
4. Unlock and open the pot. Stir in the chopped cashews before serving.

Wild Rice With Sweet Potatoes

PREP: 5 MINUTES • PRESSURE: 45 MINUTES • TOTAL: 50 MINUTES • PRESSURE LEVEL: HIGH • RELEASE: QUICK
SERVES 6

Ingredients

2 tablespoons olive oil
1 medium yellow onion, chopped
2 medium celery stalks, chopped
1 tablespoon packed fresh sage leaves, minced
2 teaspoons fresh thyme leaves
1½ cups black wild rice (about 8 ounces)
3 cups vegetable or chicken broth
1 large sweet potato (about 1 pound), peeled and diced
¼ cup dried cranberries
½ teaspoon salt
½ teaspoon ground black pepper

Directions

1. **Preparing the Ingredients.** Heat the olive oil in the COSORI Pressure Cooker Multi-Cooker® turned to the "browning" function. Add the onion and celery; cook, stirring often, until the onion softens, about 4 minutes. Mix in the sage and thyme; cook until fragrant, about 30 seconds. Stir in the rice and toss well to coat. Pour in the broth; stir well to get up any browned bits in the bottom of pot.

2. **High pressure for 30 minutes.** Lock the lid on the COSORI Pressure Cooker Multi-Cooker® and then cook for 30 minutes. To get 30-minutes cook time, press "Manual" Button.

3. **Pressure Release** Use the quick-release method to return the pot's pressure to normal. Unlock and open the cooker. Stir in the sweet potato, cranberries, salt, and pepper.

4. **High pressure for 15 minutes**. Lock the lid back on the COSORI Pressure Cooker Multi-Cooker® and then cook for 15 minutes. To get 15-minutes cook time, press "Poultry" Button and then adjust button.

5. **Pressure Release** Use the quick-release method to return the pot's pressure to normal. Unlock and open the cooker. Stir well before serving.

Brown Rice with Lentils

PREP: 5 MINUTES • PRESSURE: 35 MINUTES • TOTAL: 40 MINUTES • PRESSURE LEVEL: HIGH • RELEASE: NATURAL

SERVES 8

Ingredients

5 tablespoons olive oil
3 large onions, halved through the root (flatter) end, then sliced into thin half-moons
1 teaspoon coriander seeds
1 teaspoon cumin seeds
½ teaspoon ground turmeric
½ teaspoon ground allspice
½ teaspoon ground cinnamon
2 cups long-grain brown rice, preferably basmati
1 teaspoon sugar
1 teaspoon ground black pepper
½ teaspoon salt
4½ cups vegetable or chicken broth
½ cup green lentils (French lentils or lentils de Puy)

Directions

1. **Preparing the Ingredients.** Heat 1½ tablespoons oil in the COSORI Pressure Cooker Multi-Cooker® turned to the "Browning" function. Add half the onions and cook until well browned and crisp at the edges, at least 10 minutes, stirring occasionally. Transfer the cooked onions to a large bowl; repeat with 1½ tablespoons more oil and the rest of the onions.

   Add the remaining 2 tablespoons oil to the cooker; stir in the coriander, cumin, turmeric, allspice, and cinnamon until aromatic, about 1 minute. Add the rice, sugar, pepper, and salt; stir for 1 minute. Stir in the broth, scraping up any brown bits in the cooker. Stir in the lentils.

2. **High pressure for 35 minutes.** Lock the lid on the COSORI Pressure Cooker Multi-Cooker® and then cook for 35 minutes. To get 35-minutes cook time, press Soup button and use the TIME ADJUSTMENT button to adjust the cook time to 35 minutes.

3. **Pressure Release.** Turn off the COSORI Pressure Cooker Multi-Cooker® or unplug it so it doesn't flip to the keep-warm setting. Let its pressure return normal naturally, 14 to 20 minutes.

4. **Finish the dish** Unlock and open the cooker. Spoon the caramelized onions on top of the rice; set the lid back on the cooker without locking it in place, and set aside for 10 minutes to warm the onions. Serve by scooping up big spoonfuls with onions and rice in each.

Tiger Prawn Risotto

PREP: 10 MINUTES • PRESSURE: 30 MINUTES • TOTAL: 40 MINUTES • PRESSURE LEVEL: HIGH • RELEASE: NATURAL

SERVES 2-4

Ingredients

½ pound frozen tiger prawns, thawed and peeled
1 teaspoon salt
1 teaspoon white pepper
3 tablespoons olive oil
4 tablespoons butter
1 shallot, minced
3 cloves garlic, minced
2 cups Arborio rice
¾ cup cooking sake
2 teaspoons soy sauce
4 cups fish stock or Japanese Dashi
20 grams Parmesan cheese, finely grated
2 stalk green onions, thinly sliced

Directions

1. **Preparing the Ingredients.** In mixing bowl season the prawns with salt and white pepper. Set the COSORI Pressure Cooker Multi-Cooker® on brown and add the olive oil and butter and sauté prawns for 5-10 minutes with the shallot and garlic, the prawns should be about 80% cooked. Remove and set aside.
Add the Arborio rice, cooking sake, soy sauce and fish stock into COSORI Pressure Cooker Multi-Cooker® with a swirl of olive oil. Stir and combine, make sure the rice is coated with the liquids or Japanese Dashi

2. **High pressure for 25 minutes**. Lock the lid on the COSORI Pressure Cooker Multi-Cooker® and then cook for 25 minutes. To get 25-minutes cook time, press "Manual" button and use the TIME ADJUSTMENT button to adjust the cook time to 25 minutes.

3. **Pressure Release** Use the quick-release method to return the pot's pressure to normal. Place the prawns on top of the risotto and sprinkle the Parmesan cheese over the prawns and risotto.

4. **High pressure for 5 minutes.** Cover and lock the lid again and cook on High for another 5 minutes. To get 5-minutes cook time, press "Manual" button and use the TIME ADJUSTMENT button to adjust the cook time to 5 minutes.

5. **Pressure Release** Use the quick-release method to return the pot's pressure to normal.
Garnish with the sliced green onions.

Main Dishes – Desserts

Chocolate Pudding
White Chocolate Lemon Pudding
Blackberry Swirl Cheesecake
Poached Peach Cups with Ricotta and Honey
Molten Gingerbread Cake
Strawberry Freezer Jam Recipe
Orange Ginger Cranberry Sauce

Chocolate Pudding

PREP: 5 MINUTES • PRESSURE: 15 MINUTES • TOTAL: 20 MINUTES • PRESSURE LEVEL: HIGH • RELEASE: NATURAL

SERVES 6

Ingredients

6 ounces semisweet or bittersweet chocolate, chopped
½ ounce unsweetened chocolate, chopped
6 tablespoons sugar
1½ cups light cream
4 large egg yolks, at room temperature and whisked in a small bowl
1 tablespoon vanilla extract
¼ teaspoon salt

Directions

1. **Preparing the Ingredients.** Place all the chopped chocolate and the sugar in a large bowl. Heat the cream in a saucepan over low heat until small bubbles fizz around the inside edge of the pan.

   Pour the warmed cream over the chocolate; whisk until the chocolate has completely melted. Cool a minute or two, then whisk in the yolks, vanilla, and salt. Pour the mixture into six ½-cup heat-safe ramekins, filling each about three-quarters full. Cover each with foil.

   Set the rack in the COSORI Pressure Cooker Multi-Cooker®; pour in 2 cups water. Set the ramekins on the rack, stacking them as necessary without any one ramekin sitting directly on top of another.

2. **High pressure for 15 minutes.** Lock the lid on the COSORI Pressure Cooker Multi-Cooker® and then cook for 15 minutes. To get 15-minutes cook time, press "Manual" Button and then adjust button.

3. **Pressure Release** Turn off the COSORI Pressure Cooker Multi-Cooker® or unplug it so it doesn't flip to its keep-warm setting. Let its pressure return to normal naturally, 10 to 14 minutes.

4. **Finish the dish.** Unlock and open the cooker. Transfer the hot ramekins to a cooling rack, uncover, and cool for 10 minutes before serving—or chill in the refrigerator for up to 3 days, covering again once the puddings have chilled.

White Chocolate Lemon Pudding

PREP: 5 MINUTES • PRESSURE: 15 MINUTES • TOTAL: 20 MINUTES • PRESSURE LEVEL: HIGH • RELEASE: NATURAL

SERVES 6

Ingredients

6 ounces white chocolate, chopped

1 cup heavy cream

1 cup half-and-half

4 large egg yolks, at room temperature and whisked in a small bowl

1 tablespoon sugar

1 tablespoon finely grated lemon zest (about 1 medium lemon)

¼ teaspoon lemon extract

Directions

1. **Preparing the Ingredients.** Put the chopped white chocolate in a large bowl. Mix the cream and half-and-half in a small saucepan and warm over low heat until bubbles fizz around the edges of the pan.

   Pour the warm mixture over the white chocolate and whisk until melted. Whisk in the egg yolks, sugar, zest, and extract. Pour the mixture into six ½-cup heat-safe ramekins; cover each tightly with aluminum foil.

   Set the COSORI Pressure Cooker Multi-Cooker® rack in the COSORI Pressure Cooker Multi-Cooker®; pour in 2 cups water. Set the ramekins on the rack, stacking them as necessary without any one ramekin sitting directly on top of another.

5. **High pressure for 15 minutes.** Lock the lid on the COSORI Pressure Cooker Multi-Cooker® and then cook for 15 minutes. To get 15-minutes cook time, press "Manual" Button and then adjust button.

2. **Pressure Release.** Turn off the COSORI Pressure Cooker Multi-Cooker® or unplug it so it doesn't jump to its keep-warm setting. Let its pressure return to normal naturally, 10 to 14 minutes.

3. **Finish the dish.** Unlock and open the cooker. Transfer the (hot!) ramekins to a cooling rack; uncover each and cool for a few minutes before serving—or store in the refrigerator for up to 3 days, covering the ramekins again after they have chilled.

Blackberry Swirl Cheesecake

PREP: 10 MINUTES • PRESSURE: 20 MINUTES • TOTAL: 30 MINUTES • PRESSURE LEVEL: HIGH • RELEASE: NATURAL

SERVES 4-6

Ingredients

1 cup fresh blackberries
½ cup powdered sugar
4 tablespoons unsalted butter
1 cup crushed graham crackers
14 ounces cream cheese (one 8-ounce and two 3-ounce packages)
½ cup granulated sugar
Freshly grated zest from 1 lemon
Freshly grated zest from half an orange
2 large eggs

Directions

1. **Preparing the Ingredients.** Add 2 cups of water to the COSORI Pressure Cooker Multi-Cooker® base; insert the steamer basket and set aside. Cut a piece of wax paper to fit the bottom of a wide, flat-bottomed 4-cup baking dish; also cut a strip sized to fit the sides of the dish. Line the dish with the paper.

   Puree the blackberries and powdered sugar in a blender and set aside.

   Melt the butter in a medium saucepan on medium heat. Remove the pan from the heat and mix in the crushed crackers. Scoop the mixture into the prepared baking dish and, using the back of your hand, push it into a flat, thin, even layer that covers the bottom of the dish, and, if there is enough, partway up the sides. Put the dish in the refrigerator to chill, uncovered, while you prepare the filling.

   In a medium bowl, using an electric mixer on medium speed, mix together the cream cheese, granulated sugar, and lemon and orange zests. Add the eggs and mix into a smooth batter, about 5 minutes.

   Remove the dish with the crust from the refrigerator. Slowly pour the batter over the crust, spreading level. To add the blackberry swirl, pour the puree into a squirt bottle (or food storage bag with one corner clipped off) and with it draw a spiral from the center out on top of the batter. Then use a toothpick or skewer to drag radiating lines from the center to the edge of the dish. Using a foil sling, lower the dish into the COSORI Pressure Cooker Multi-Cooker®; do not cover the dish.

2. **High pressure for 20 minutes.** Lock the lid on the COSORI Pressure Cooker Multi-Cooker® and then cook for 20 minutes. To get 20-minutes cook time, press "Manual" button and use the TIME ADJUSTMENT button to adjust the cook time to 20 minutes.

3. **Pressure Release.** When the time is up, open the COSORI Pressure Cooker Multi-Cooker® using the 10-Minute Natural Release method.

4. **Finish the dish.** Lift the dish out of the COSORI Pressure Cooker Multi-Cooker® and check the cake for doneness, transfer the dish to a wire rack.

Let the cake cool, uncovered, for about 30 minutes. Then cover the dish with plastic wrap and refrigerate until ready to serve, for at least 4 hours.

Work quickly and delicately to unmold the chilled cake: Invert a plate over the dish and flip the dish and plate over together. Lift the dish off the cake and then peel off the wax paper circle on the base and the strip on the sides. Then invert a serving plate on the cake and gently flip all three components over together; lift off the top plate. Serve the cake cold, cut into wedges.

Poached Peach Cups with Ricotta and Honey

PREP: 5 MINUTES • PRESSURE: 4 MINUTES • TOTAL: 9 MINUTES • PRESSURE LEVEL: LOW • RELEASE: QUICK
SERVES 4

Ingredients

    4 peaches, cut in half and pitted
    1/4 cup apple juice
    1/4 cup water
    3 tablespoons light brown sugar
    1/8 teaspoon ground cinnamon
    1 cup part-skim ricotta cheese
    2 tablespoons honey
    1/4 teaspoon vanilla extract

Directions

1. **Preparing the Ingredients.** Add peaches, apple juice, water, brown sugar, and cinnamon to the cooker.
2. **High pressure for 5 minutes.** Lock the lid on the COSORI Pressure Cooker Multi-Cooker® and then cook for 5 minutes. To get 5-minutes cook time, press "Manual" button, and use the TIME ADJUSTMENT button to adjust the cook time to 5 minutes.
3. **Pressure Release.** Perform a quick release to release the cooker's pressure. Remove peaches from cooking liquid, and set aside.
4. **Finish the dish.** Combine ricotta cheese, honey, and vanilla extract, and serve spooned into the center of each peach half.

Molten Gingerbread Cake

PREP: 5 MINUTES • PRESSURE: 15 MINUTES • TOTAL: 20 MINUTES • PRESSURE LEVEL: HIGH • RELEASE: COMBINATION

SERVES 2

Ingredients

3 tablespoons very hot water
¼ cup vegetable oil
¼ cup packed brown sugar
¼ cup molasses
1 large egg
⅔ cup all-purpose flour
¾ teaspoon ground ginger
½ teaspoon ground cinnamon
¼ teaspoon kosher salt
¼ teaspoon baking powder
¼ teaspoon baking soda
1 cup water, for steaming (double-check the COSORI Pressure Cooker Multi-Cooker® manual to confirm amount, and follow the manual if there is a discrepancy)

Directions

1. **Preparing the Ingredients.** In a small bowl, using a hand mixer, mix together the hot water, vegetable oil, brown sugar, molasses, and egg. In another small bowl, sift together the flour, ground ginger, cinnamon, kosher salt, baking powder, and baking soda. Add the dry ingredients to the liquid mixture. Mix on medium speed until the ingredients are thoroughly combined, with no lumps. Pour the batter into a nonstick mini (3-by-5-inch) loaf pan. Cover the pan with aluminum foil, making a dome over the pan.
Add the water and insert the steamer basket or trivet. Carefully place the loaf pan on the steamer insert.
2. **High pressure for 15 minutes**. Lock the lid on the COSORI Pressure Cooker Multi-Cooker® and then cook for 15 minutes. To get 15-minutes cook time, press "Manual" button and use the TIME ADJUSTMENT button to adjust the cook time to15 minutes. When the timer goes off, turn the cooker off. ("Warm" setting, turn off).
3. **Pressure Release.** After cooking, use the natural method to release pressure for 5 minutes, then the quick method to release the remaining pressure.
4. **Finish the dish.** Unlock and remove the lid. Using tongs, carefully remove the pan from the COSORI Pressure Cooker Multi-Cooker®. Let the cake rest for 2 to 3 minutes; remove the foil, slice, and serve.
 To sift the dry ingredients, place a medium-coarse sieve over a small bowl or on a sheet of wax paper or parchment paper. Measure the dry ingredients into the sieve. Tap the side of the sieve to move the contents through the sieve to the bowl or parchment paper; then transfer the sifted ingredients to the wet ingredients.

PER SERVING: CALORIES: 639; FAT: 30G; SODIUM: 506MG; CARBOHYDRATES: 87G; FIBER: 2G; PROTEIN: 8G

Strawberry Freezer Jam Recipe

PREP: 10 MINUTES • PRESSURE: 8 MINUTES • TOTAL: 18 MINUTES • PRESSURE LEVEL: HIGH • RELEASE: NATURAL
SERVES 4 CUPS

Ingredients

1 lb. (450 grams) strawberries (fresh or frozen)
1 /2 to 1 lb. (225 to 450 grams) granulated sugar
1 navel orange
1 tbs. butter (optional, vegans can omit)

Directions

1. **Preparing the Ingredients.** If you are using fresh berries, remove the stems, leaves and any bruised spots from the strawberries, lightly wash them, and cut into halves or quarters, depending on size. For frozen berries, defrost before use, cut them up if necessary.

   Peel the navel orange, removing the bitter white pith and any white connective tissues. I do this by slicing a bit off the top, so I can see how thick the peel is. I then take the knife and cut slices of peel down the sides of the orange. (It is better to remove a little of the orange than to leave the bitter pith on the orange.) Once you have removed all the peel and any pith attached to the outside of the orange, break it apart into segments, remove any white pithy connective tissues inside, and roughly chop the segments. Reserve the chopped segments and any juice.

   For a very smooth jam, place the sliced strawberries and chopped orange segments and juice into a food processor or blender and puree until smooth, then add to the sugar. If you would like your jam more like preserves (with small pieces of fruit mixed in), combine the sliced strawberries, orange pieces, and orange juice into the sugar.

   Once mixed, use a potato masher to roughly mash the strawberries. The mixture should macerate in the refrigerator for at least an hour, but if you can let it set for 8 – 24 hours, that's even better.

   Once the mixture has macerated, add to the COSORI Pressure Cooker Multi-Cooker®. Using the "Browning" setting, bring the jam up to a hard boil for 3 minutes to dissolve the sugar and reduce the excess water content. Stir frequently with the longest handled spatula you own.

   Stir in 1 tablespoon of butter

2. **High pressure for 8 minutes.** Lock the lid on the COSORI Pressure Cooker Multi-Cooker® and then cook for 8 minutes. To get 8-minutes cook time, press "Manual" button, and use the TIME ADJUSTMENT button to adjust the cook time to 8 minutes.

3. **Pressure Release.** Let its pressure return to normal naturally, 8 to 12 minutes.

4. **Finish the dish.** After pressure has released, unlock and remove the lid, tilting the front side down and the back side up to direct any residual heat and steam away from you. With the lid OFF and the "Browning" setting, bring the mixture back to the boil for 3 minutes, stirring frequently. Turn the unit off after the 3 minutes are up. Allow mixture to cool to room temperature, stirring periodically. Once cooled, put the jam in a container in the refrigerator to finish setting.

Orange Ginger Cranberry Sauce

PREP: 5 MINUTES • PRESSURE: 4 MINUTES • TOTAL: 9 MINUTES • PRESSURE LEVEL: HIGH • RELEASE: QUICK
SERVES 8

Ingredients

1 pound Fresh Cranberries
1 cup Orange Juice
½ cup Demerara Sugar, maple syrup, or sugar, to taste
1 teaspoon Cinnamon, ground
1 teaspoon True Orange Ginger
or 1 teaspoon Fresh Ginger, grated (optional)
1 Cinnamon Stick
Zest from one Large Orange

Directions

1. **Preparing the Ingredients.** Rinse and drain cranberries and set aside.
   Whisk together orange juice, sugar, cinnamon and True Orange Ginger (or grated ginger).
   Add cranberries and juice to the COSORI Pressure Cooker Multi-Cooker® cooking pot.
   Add cinnamon stick.
2. **High pressure for 4 minutes.** Lock the lid on the COSORI Pressure Cooker Multi-Cooker® and then cook for 4 minutes. To get 4-minutes cook time, press "Manual" button and use the TIME ADJUSTMENT button to adjust the cook time to 4 minutes.
3. **Pressure Release.** When beep sounds, allow a full Natural Pressure Release.
   For a thicker Sauce, simmer until desired consistency.
   Open the lid and add Orange Zest.

HERITAGE OF FOOD: A FAMILY GATHERING

To survive, we need to eat. As a result, food has turned into a symbol of loving, nurturing and sharing with one another. Recording, collecting, sharing and remembering the recipes that have been passed to you by your family is a great way to immortalize and honor your family. It is these traditions that carve out your individual personality. You will not just be honoring your family tradition by cooking these recipes but they will also inspire you to create your own variations, which you can then pass on to your children's.

The recipes are just passed on by everyone and nobody actually possesses them. I too love sharing recipes. The collection is vibrant and rich as a number of home cooks have offered their inputs to ensure that all of us can cook delicious meals at our home. I am thankful to each one of you who has contributed to this book and has allowed their traditions to pass on and grow with others. You guys are really wonderful!

I am also thankful to the cooks who have evaluated all these recipes. You're, as well as, the comments that came from your family members and friends were really invaluable.

If you have the time and inclination, please consider leaving a short review wherever you can, we would love to learn more about your opinion.

https://www.amazon.com/review/review-your-purchases/

Resources

http://instantpot.com
http://dadcooksdinner.com/2015/01/instant-pot-frequently-asked-questions.html/
http://www.pressurecookrecipes.com/instant-pot-water-test/
https://instantpot.com/wp-content/uploads/IP-LUX-v2/InstantPot-IP-LUX-Manual-English-v2.pdf
http://www.pressurecookingtoday.com/how-to-convert-a-recipe-into-a-pressure-cooker-recipe/
www.ext.colostate.edu/pubs/columncc/cc970123.htm.
http://instantpot.com/cooking-time/
http://www.convert-me.com/en/convert/cooking/
http://instantpot.com/faq/
http://woodhavenpl.com/instant-pot-freezer-meals-tutorial/
http://instantpot.com/benefits/specifications-and-manuals/

http://www.pressurecookingtoday.com/how-to-use-the-power-pressure-cooker-xl

About the Author

Betty is a Chicago-based food writer, experienced chef. She has contributed food articles in Magazines and Blogs, growing out of her commitment to make it possible for everyone to cook, even if they have too little time. She loves sharing Easy, Delicious and Healthy recipes, especially with those who are brand-new to the Pressure Cooker. Making the family dinner was one of her weekly chores, and she loved inventing crazy recipes with her best friend for their afternoon snacks. When she's not cooking, Janie enjoys spending time with her husband and her kids, gardening and traveling.

CPSIA information can be obtained
at www.ICGtesting.com
Printed in the USA
LVOW09s2052070118
562153LV00017B/276/P